T0351037

Electromagnetic Bandgap (EBG) Structures

Electromagnetic Bandgap (EBG) Structures

Common Mode Filters for High-Speed Digital Systems

Antonio Orlandi, Bruce Archambeault,
Francesco de Paulis, and Samuel Connor

IEEE PRESS

WILEY

Published by John Wiley & Sons, Inc., Hoboken, New Jersey.
Published simultaneously in Canada.

For general information on our other products and services or for technical support, please contact our Customer Care Department within the United States at (800) 762-2974, outside the United States at (317) 572-3993 or fax (317) 572-4002.

Wiley also publishes its books in a variety of electronic formats. Some content that appears in print may not be available in electronic formats. For more information about Wiley products, visit our web site at www.wiley.com.

Library of Congress Cataloging-in-Publication Data is available.

ISBN: 978-1-119-28152-8

Printed in the United States of America.

10 9 8 7 6 5 4 3 2 1

Contents

About the Authors

Antonio Orlandi (M'90–SM'97–F'07) was born in Milan, Italy in 1963. He received the Laurea degree in Electrical Engineering from the University of Rome "La Sapienza," Italy, in 1988 and the Ph.D. degree in Biomedical Engineering from the University "Campus Biomedico," Italy, in 2012. He was with the Department of Electrical Engineering, University of Rome "La Sapienza" from 1988 to 1990. Since 1990, he has been associated with the Department of Electrical Engineering of the University of L'Aquila where he is currently Full Professor and Chair of the *UAq EMC Laboratory*. Author of more than 320 technical papers, he has published in the field of electromagnetic compatibility in lightning protection systems and power drive systems. Current research interests are in the field of numerical methods and modeling techniques to approach signal/power integrity, EMC/EMI issues in high-speed digital systems. Dr. Orlandi received the IEEE Transactions on Electromagnetic Compatibility Best Paper Award in 1997, IEEE Transactions on Advanced Packaging Best Paper Award in 2011, the IEEE EMC Society Technical Achievement Award in 2003 and 2012, the IBM Shared University Research Award in 2004, 2005, 2006 and 2012, the CST University Award in 2004, the IEEE International Symposium on EMC Best Paper Award in 2009, 2010, 2013, the IEEE International Conference on SI/PI Best Paper Award in 2016, the DesignCon Best Paper Award in 2011 and 2012, and the IEEE Transactions on Electromagnetic Compatibility Best Paper Award Honorable Mention in 2015. He is co-recipient of the 2015 CISCO University Program Fund Award on "TSV modeling and measurement." From 1996 to 2000, and from 2010 up to December 2015, he has been Associate Editor of the *IEEE Transactions on Electromagnetic Compatibility*. Since January 2016, he is the Editor-in-Chief of the *IEEE Transactions on Electromagnetic Compatibility*. Dr. Orlandi is member of the "Education," TC-9 "Computational Electromagnetics," and Past Chairman of the TC-10 "Signal Integrity" Committees and Vice-Chair of TC-12 "EMC for Emerging Wireless Technologies" of the IEEE

EMC Society. He is General Co-Chair of the 2017 IEEE SIPI Conference. From 2001 to 2006, he served as Associate Editor of the *IEEE Transactions on Mobile Computing* and from 1999 to the end of the Symposium was Chairman of the TC-5 "Signal Integrity" Technical Committee of the International Zurich Symposium and Technical Exhibition on EMC.

Bruce Archambeault is an IEEE Fellow, an IBM Distinguished Engineer Emeritus, and an Adjunct Professor at Missouri University of Science and Technology. He received his B.S.E.E degree from the University of New Hampshire in 1977 and M.S.E.E degree from Northeastern University in 1981. He received his Ph.D. from the University of New Hampshire in 1997. His doctoral research was in the area of computational electromagnetics applied to real-world EMC problems. He has taught numerous seminars on EMC and Signal Integrity across the United States and the world, including the past 15 years at Oxford University.

Dr. Archambeault has authored or coauthored a number of papers in computational electromagnetics, mostly applied to real-world EMC applications. He is a member of the Board of Directors for the IEEE EMC Society and a past Board of Directors member for the Applied Computational Electromagnetics Society (ACES). He currently serves as the President-Elect for Conferences of the EMC Society. He has served as a past IEEE/EMCS Distinguished Lecturer, EMCS TAC Chair, and Associate Editor for the *IEEE Transactions on Electromagnetic Compatibility*. He is the author of the book *PCB Design for Real-World EMI Control* and the lead author of the book titled *EMI/EMC Computational Modeling Handbook*.

Francesco de Paulis (S'08, M'13) was born in L'Aquila, Italy in 1981. He received the Specialist degree (summa cum laude) in Electronic Engineering from University of L'Aquila, L'Aquila, Italy, in 2006. In August 2006, he joined the EMC Laboratory at the Missouri University of Science and Technology (formerly University of Missouri-Rolla), USA, where he received the M.S. degree in Electrical Engineering in May 2008. He received the Ph.D. degree in Electrical and Information Engineering in 2012 from the University of L'Aquila, L'Aquila, Italy.

He was involved in the research activities at the UAq EMC Laboratory, L'Aquila, Italy, from August 2004 to August 2006 and at the MST EMC Laboratory, Rolla, MO, from August 2006 to May 2008. From June 2004 to June 2005, he had an internship at Selex Communications, L'Aquila, within the layout/SI/PI design group. He is currently a Research Assistant at the UAq EMC Laboratory, University of L'Aquila, Italy. His main research interests are in developing fast and efficient analysis techniques for SI/PI and design of high-speed signals on PCB and packages, analysis and characterization of composite materials for shielding, RF interference in mixed-signal system, TSVs in silicon chips and interposers, EMI problem investigation, TDR techniques, and fault and degraded joint remote detection in power transmission lines.

Dr. de Paulis received from the IEEE EMC Society the Past President's Memorial Award in 2010. He was the recipient of the Best Paper Award in 2010, 2013, and 2016, and the Best Student Paper Award in 2009 and in 2011 at the IEEE International Symposium on EMC. He received the Paper Award in the power and RF design category in 2010, 2011, and 2012 at the IEC DesignCon. He was selected as Distinguished Reviewer of the Transaction on EMC for the year 2014.

Samuel Connor (M'04–SM'07) received his Bachelor's Degree in Electrical Engineering from the University of Notre Dame in 1994. Sam is currently a Senior Technical Staff Member in the IBM Systems Group, where he leads the EMC Design Center of Competency. Sam is a past-chair for both TC-9 and the Eastern North Carolina Chapter of the IEEE EMC Society, has served on numerous EMC Symposium Committees, and was a Distinguished Lecturer during 2012–2013.

Preface

As differential digital signal data rates increased to many Gb/s range, it became apparent that small amounts of mismatch within a differential pair of traces could create common mode signals that could adversely affect the electromagnetic compatibility (EMC) performance of systems. Therefore, it became important to find a way to filter the unwanted common mode signals while not affecting the signal integrity of the differential signals. Unfortunately, discrete common mode filter components can be used to only a few Gb/s data rate without too much impact on the intentional differential signal, so an alternative was needed.

Electromagnetic bandgap (EBG) filters appeared like a perfect solution, since no discrete components are required. However, at the time this research effort began, the previous work on EBGs was more of a "try it and see what we get" approach rather than a straightforward design that could be used in the real world of high-speed data communications product development. The planar EBG technology, deeply studied by the same authors and initially conceived for noise mitigation in power distribution networks, was found appropriate to develop resonance-based common mode filters.

This research to develop straightforward design equations to predict the required size of the EBG elements and to validate these equations with simulations and measurements has taken a number of years and a team approach. The primary researchers on the team are from the University of L'Aquila, but team members from IBM and the Missouri University of Science and Technology also contributed throughout the effort. This team approach of university and industry has worked well to create a design approach that is useful in the real world, while still being interesting and worthy of academia! The result is a set of filter designs that maintain reasonable insertion loss and cross talk performance of the intentional differential signal, strike a good balance between miniaturization and design complexity, and provide alternatives for mitigating the direct radiation from the filter structure.

The authors want to acknowledge the contributions of Dr. Carlo Olivieri at University of L'Aquila as well as Dr. Michael Cracraft of IBM. Their contributions to this work have been significant!

Acknowledgments

To my wife Antonia and my "kids" Michele, Anna, and Cecilia.
—Antonio Orlandi

Dedicated to my wife of 45 years for her constant support and love.
—Bruce Archambeault

I would like to express my gratitude to all researchers and engineers who contributed, throughout the last decade, to the development of the EBG project, and who do not appear explicitly in this book. In particular, Leo Raimondo and Danilo Di Febo of the UAq EMC Laboratory (University of L'Aquila), Muhammet Hilmi Nisanci of the UAq EMC Laboratory, and now of Sakarya University (Turkey), Eng. Riccardo Cecchetti of Technolabs/Intecs (L'Aquila plant), Dr. Jun Fan of the MS&T EMC Laboratory (Missouri University of Science and Technology), and Dr. Xiaoxiong Gu of IBM.
—Francesco de Paulis

Many thanks to my IBM colleagues who contributed to the design, fabrication, and testing of the EBG test boards over the course of our research. And most of all, I would like to thank my wife, Caroline, for her love and support throughout this project.
—Samuel Connor

1

Introduction

1.1 Motivations

This book is focused on a specific use of electromagnetic bandgap (EBG) structures: their function as common-mode (CM) filter in high-speed differential digital systems and/or hybrid mixed-signal circuits.

In order to appreciate the potential of these structures as signal filter, it is instructive to give a look to the historical development of the EBG structures at least since 1999 when they were proposed as high-impedance electromagnetic surfaces for band-stop filter [1,2].

The first application was related to flat metal sheets used in many antennas as reflectors or ground planes. These sheets support surface waves [3,4], that is, propagating electromagnetic waves that are bound to the interface between metal and free space. If the metal surface is smooth and flat, the surface wave will not couple to external propagating plane waves. However, they will radiate vertically if scattered by bends, discontinuities, or surface texture and this can generate, in case of multiple antenna placement, unwanted mutual coupling and interference.

By applying a special texture on a conducting surface, it is possible to alter its electromagnetic properties [5,6]. In the limit where the period of the surface texture is much smaller than the wavelength, the structure can be described using an effective medium model, and its qualities can be described by the surface impedance. A smooth conducting sheet has a low surface impedance; however, with a specially designed textured surface, the sheet can have a high surface impedance, thus inhibiting the flow of the currents over a selected frequency range.

The first example of EBG as high-impedance surface consisted in an array of metal protrusions on a flat metal sheet. They are arranged in a two-dimensional lattice and can be visualized as mushrooms protruding from the surface [7–11]. The surface can be easily fabricated using standard printed circuit boards (PCB)

Electromagnetic Bandgap (EBG) Structures: Common Mode Filters for High-Speed Digital Systems,
First Edition. Antonio Orlandi, Bruce Archambeault, Francesco De Paulis, and Samuel Connor.
© 2017 by The Institute of Electrical and Electronics Eingineers, Inc. Published 2017 by John Wiley & Sons, Inc.

technology. The protrusion are formed as metal patches connected to the lower continuous conducting surfaces by plated through-hole vias.

If the protrusions are small compared to the operating wavelength, their electromagnetic behavior can be described by using the lumped circuit theory. The EBG structure behaves like a network of parallel resonant LC circuits, which act as a two-dimensional electric filter to block the flow of currents along the sheet. In the frequency range where the surface impedance is high, the tangential magnetic field is small, even with a large electric field along the surface.

The mushroom-type EBG configuration has inspired the PCB designers to use this structure for suppressing noise in power planes [12]. An ideal power delivery network (PDN) is assumed to supply clean power to integrated circuits. However, electromagnetic noise in power/ground-reference planes can cause fluctuation or disturbance in the power supply voltage, which, in turn, leads to false switching, jitter, and malfunctioning in analog or digital circuits. Modern digital electronic circuits have increased the clock frequency and pulse edge rate, and has contributed to the decreased of the power supply voltage and noise margin. This power/ground-reference noise creates significant challenges for electromagnetic compatibility and signal/power integrity engineers. Simultaneous switching noise has become one of the major concerns [13,14] in high-speed PCB design.

This type of disturbance has been discussed extensively over the last decade [15–21] and different approaches have been proposed. Most prominent of these involve the use of discrete decoupling capacitors and embedded capacitances [22,23]. However, this approach fails when operated at high frequencies due to the inherent inductance of discrete capacitors and especially the inductance associated with connecting the capacitors to the power/ground-reference planes. Embedded capacitance is usually two very closely spaced planes (often with a higher than normal dielectric constant); it is an expensive solution and reliability considerations limit its practical use. Mushroom EBGs have proven effective for noise suppression at frequencies above 1 GHz and can be effective when discrete capacitors and/or embedded capacitance cannot be effective. When the mushroom-type configuration is implemented in PCB, it uses three layers where the EBG pattern layer with specially designed vias is inserted between the power plane and a ground-reference plane, as shown in Figure 1.1. This configuration makes the fabrication more expensive since extra PCB layers are used for the filter.

The natural evolution of the mushroom-type EBG applied to PDN in printed circuit boards have been the planar EBG structures used either for switching noise mitigation or in mixed-signal boards [24–33]. These structures consist of a power distribution system of only two layers, instead of three of the mushroom type, with one of the layers patterned in a periodic fashion, effectively creating a frequency band-stop filter. These structures, in contrast

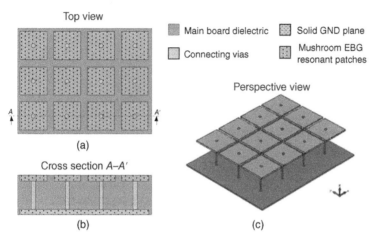

Figure 1.1 Mushroom EBG configuration. (a) Top view. (b) Cross-sectional view. (c) Perspective view.

to the previously described mushroom filters, do not have vias or require the third layer. These features make such structures very attractive for PCB applications from the manufacturing and cost perspectives.

Their basic structure is illustrated in Figure 1.2.

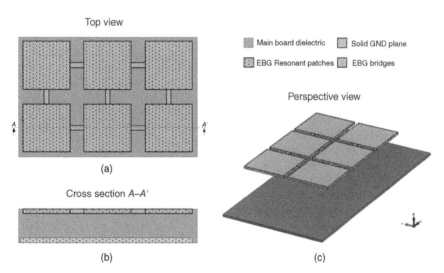

Figure 1.2 Planar EBG configuration. (a) Top view. (b) Cross-sectional view. (c) Perspective view.

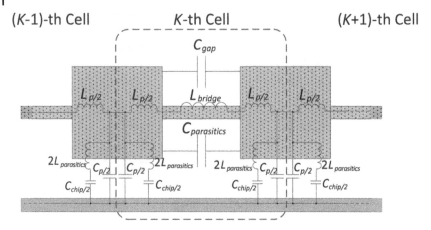

Figure 1.3 Qualitative equivalent circuit of a unit cell of a planar EBG structure.

In this basic structure, the solid layer can be used for one voltage level and the EBG patterned one for a second voltage level (often ground-reference). Between these two layers, there is a uniform substrate material whose nature (organic, ceramic, lossy, etc.) depends on the application of the board and the performances of the filter. For one-dimensional wave propagation, the unit cell of this planar EBG structure can be modeled with the basic equivalent circuit shown in Figure 1.3 [34–38].

The left part of the figure describes the propagation characteristics between the EBG patch and the continuous power plane represented by the equivalent patch inductance L_p and capacitance C_p. The second part of the figure characterizes the bridge effects between two adjacent unit cells. The gap between two patches generates a fringing electric filed associated with the equivalent capacitance C_b and the bridge's inductance L_{bridge}. A repetition of these cells can be conceptually viewed as an electric filter of parallel LC resonators.

The basic structure of the mushroom-like EBG structure has evolved to the concept of the ground-reference surface perturbation lattice (GSPL) geometry [39–44]. This structure is similar to the EBG filter but with multiple vias, and its design or use is typically appropriate when there is a need to enhance the bandwidth of the bandgap for power delivery noise suppression [45–49]. By using multiple shorting vias and optimizing their arrangement, the GSPL structure presents a wider bandwidth bandgap than that of the mushroom-like structure. In the GSPL, the mechanism of the bandwidth enhancement is based on the optimization of the vias locations. A one-dimensional equivalent circuit model, conceptually similar to that illustrated in Figure 1.3, can be used to predict the stopband. Test structures are manufactured on FR4 substrate to

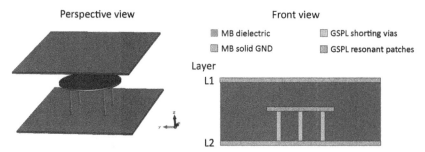

Perspective view   Front view

■ MB dielectric   ▢ GSPL shorting vias
▢ MB solid GND   ▒ GSPL resonant patches

Figure 1.4 GSPL with four vias. MB = mother board.

compare the measured results and the numerical ones. Figure 1.4 shows a GSPL with four vias.

After the previous brief review of the main frequency-selective structures similar to or derived from the EBGs, it is possible to move toward the description of a more specific application: their use as signals filter in digital systems.

Where data rates get into the high hundreds of megabits (Mb/s) or gigabits (Gb/s), signal integrity (SI) concerns will usually require that differential signaling is used in order to ensure the required signal quality. Dielectric loss for long traces, reflections from connectors and vias, and even surface roughness will reduce signal quality at the end of long traces at very high data rates.

Differential signaling is also more immune to external noise corrupting the intentional signals. The basic intention for differential signals is for two equal and opposite currents (and voltages) to exist on the pair of traces, and the ground-reference plane plays no role in the intentional signal current. In reality, this is true only when there are only two signal conductors in free space, with no other metal nearby. This perfect condition never occurs in typical printed circuit boards [50]; therefore, there is always some RF currents on the ground-reference plane in real-world PCBs.

The presence of common-mode noise in the differential signal is one of the main causes of electromagnetic interference (EMI) problems in chip packages and printed circuit boards, especially in the gigahertz range of state-of-the-art high-speed digital systems. The common-mode signal can propagate outside the shielded enclosure through connectors and cables and cause unwanted external radiation.

The previously introduced EBG structures are primarily used for noise mitigation in PCBs and packages, thus enhancing the power integrity performance of the power delivery network [51,52]. The regular planar EBG is investigated in Ref. [53], studying the effects of the patterned plane on both common-mode and differential-mode signal propagation along a differential

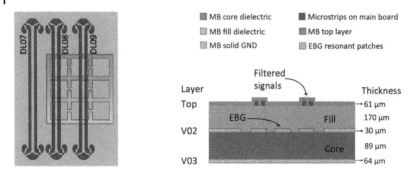

Figure 1.5 Basic onboard EBG CM filter structure for common-mode filtering: external layers layout.

microstrip line. These principles are applied for filtering the common-mode noise (due to some imbalance) in a differential signal [54–65].

The electromagnetic properties and the layout technique regulating the EBG common-mode filter behavior will be discussed in several parts and under different perspectives across the book. These EBG filters can be placed near I/O connectors on PCBs to reduce the amount of common-mode current that is coupled onto the cables or near ICs to suppress the common-mode noise near its source.

The most simple EBG-based CM filter is laid out on the PCB outermost stack-up layer (the so-called top and bottom layers) as in Figure 1.5 and is sometime referenced as an onboard EBG CM filter. The figure shows the real layout of a manufactured board that was employed for investigating the cross talk among adjacent differential pairs routed on the same EBG filter [66,67].

These onboard EBG CM filters can also be laid out on the internal layers of the stack-up, as shown in Figure 1.6. The stripline filter consists of two patterned layers above and below the differential traces: In this way the return current flows on both the planes above and underneath the traces. A possible variation to the classic EBG structure is the removing of the bridges connecting the patches. This new configuration (Figure 1.7) in general provides deeper notches (but less bandwidth of the bandgap filter) than the regular EBG structure for filtering the common-mode signal.

These EBG filter configurations are designed to attenuate the common-mode component of the signal, as shown by the common-mode mixed-mode scattering parameter S_{cc21} in Figure 1.8a, without affecting the transmission of the differential mode and thus without spoiling signal integrity of the output eye diagram as shown in Figure 1.8b.

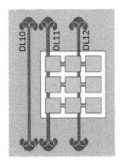
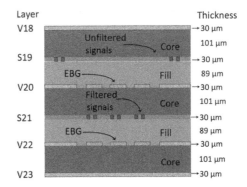

Figure 1.6 Basic onboard EBG CM filter structure for common-mode filtering: internal layers layout.

A different layout strategy was adopted in Refs [68–73] to provide more flexibility in the filter design. The EBG filter is eliminated from the PCB stack-up, and it is modified to be a surface-mount component installed on top of a PCB. In the literature, this configuration is referred to as a *removable* EBG CM filter. Also, with this configuration, the key design concepts such as the use of standard multilayer laminate technology, the straightforward design procedure, and the reduced costs that make the EBG filter attractive are still valid. Moreover, the electromagnetic behavior of the filter remains unchanged,

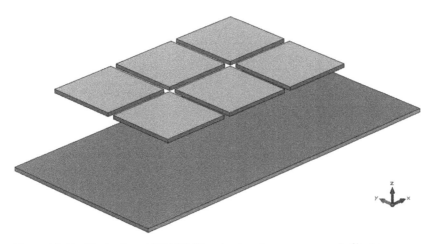

Figure 1.7 Modified onboard EBG CM filter structure for common-mode filtering.

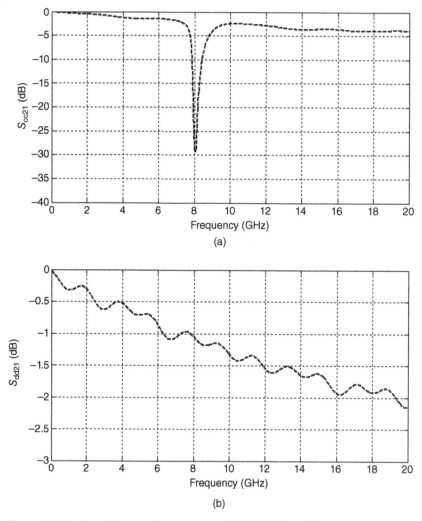

Figure 1.8 Mixed-mode scattering parameters for EBG CM filters. (a) S_{cc21}. (b) S_{dd21}.

with the common-mode return currents of the differential pair being responsible for the common mode to EBG cavity mode coupling. The PCB area used by the removable EBG CM filter can be minimized by employing techniques for its miniaturization; the simplest strategy is to utilize a high-permittivity material whose larger costs, with respect to the standard laminates (i.e., FR-4), remains limited to a millimeter-size multilayer PCB rather than the main PCB.

A qualitative example of removable EBG CM filter is given in Figure 1.9.

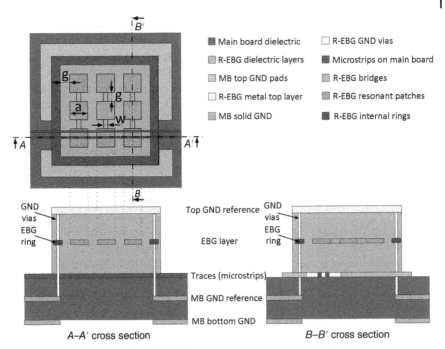

Figure 1.9 Basic removable EBG CM filter structure for common-mode filtering.

The filter is attached to the PCB by means of four corner pads for the current return corresponding to pads on the PCB.

The performances of such *removable* configuration are as good as the *onboard* counterpart. Figure 1.10 shows a S_{cc21} for this filter, very well centered

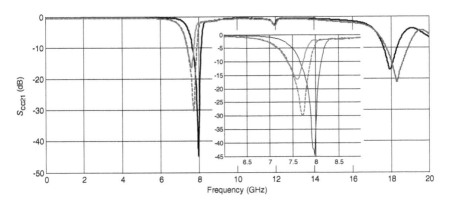

Figure 1.10 Mixed-mode scattering parameters S_{cc21} for a removable EBG CM filter.

on the design frequency (i.e., 8 GHz) and its sensitivity due to the variation of geometrical parameters, as will be discussed in this book.

In conclusion of this brief introduction to the topic of the implementation of the electromagnetic bandgap structures as CM filters for high-speed differential signals, it should be mentioned that the actual research trend is toward the miniaturization of these structures in order to minimize their dimensions without affecting the filtering performances [74].

The more recent scientific literature shows two different approaches to reach this goal: the use of material with high dielectric permittivity and the design of patterned structures using novel resonators with limited dimensions to excite the filter resonances.

The former approach has explored the use of ceramic dielectric such as the low-temperature co-fired ceramic (LTCC) [75,76] and it is suitable for the use of the removable filters because they allow a decoupling between the dielectric material of the main board and that of the component itself.

The latter is showing very promising miniaturization factors of around 10 times the standard EBG CM filters [77–79]. These resonators can be easily implemented on PCB or even on package substrate by designing an open stub with shorting via connecting to the reference plane. This configuration provides the shorting path of the common-mode return currents at gigahertz range and still maintains the isolation at DC level.

Finally, particularly significant to have a complete outlook of the EBG field of applications are Refs [80–83].

All the three-dimensional full-wave simulations have been performed by using the *CST Studio Suite 2015* by *Computer Simulation Technology* (CST) [84] and the *Advanced Design System* (ADS) environment by *Keysight Technology* [85] for the transient and frequency analysis of the equivalent circuit models.

1.2 Scope of the Book

The book aims at providing the basic principles of operation of the planar EBG structures as common-mode filters for high data rate digital systems. The following is a brief description of the chapterwise coverage of different topics.

This chapter introduces the topic of the book, offering a brief historical perspective of the introduction and use of the EBGs in the printed circuit board world and their evolution into CM filters.

Chapter 2 describes the fundamental mechanisms of planar EBGs looking into details of the mechanisms of resonances and the definition of the lower and higher boundaries of the bandgap as well as proposes the design criteria for these structures with particular emphasis on their impact on power integrity.

Chapter 3 is devoted to the study of the structures described in Chapter 2, but also looking at their impact on the integrity of signals flowing on single-ended and/or differential traces routed above or between EBG filters. This chapter shows the common-mode filter on a differential trace referenced to a patterned plane is equivalent to the response of a single-ended trace reference to the same plane. This finding will be the basis for the use of the EBGs as filters.

Chapter 4 introduces the concept of *onboard* EBG CM filter based on a simple patch resonant cavity. This approach permits a simple and detailed theoretical treatment that allows the reader to easily design their own EBG CM filter for their specific application. Some full-wave examples and simulation results are presented and compared to validate the design approach.

Chapter 5 contains few specific topics concerning the design and implementation of EBG CM filters:

• Techniques to enhance the bandwidth of the bandgap associated with the EBG such as multiple size of patches and bridges.
• Approaches to reduce the overall size of the EBG on the printed circuit board.

Chapter 6 is similar in structure to Chapter 4: It discusses the evolution of the *onboard* EBG CM filters in *removable* EBG CM filters. These removable filters allow designers to replace EBG filters on the board according to their needs without redesigning the overall board. This chapter presents different topologies of removable EBG CM filters, from the one with the traces kept on the main PCB to the second configuration with the differential pair routed on (or inside) the removable part. This chapter also contains details of the miniaturization techniques for the EBG CM filters and their external electromagnetic radiation.

Chapter 7 describes a number of measurements made to validate the operations of CM EBG filters as designed in previous chapters. The main content of this chapter is to provide information and details of the measurement setup and procedures to measure the signal integrity and EMC performances of these EBG structures. A detailed description of the measurement techniques and of the calibration and de-embedding strategies are included.

References

1 D. Sievenpiper, *High-impedance electromagnetic surfaces*. Ph.D. dissertation, Department of Electrical Engineering, University of California at Los Angeles, CA, 1999.

2 D. Sievenpiper, L. Zhang, R.F.J. Broas, N.G. Alexopulos, and E. Yablonovitch, High-impedance electromagnetic surfaces with a forbidden frequency band. *IEEE Trans. Microw. Theory Tech.*, Vol. 47, No. 11, 1999, pp. 2059–2073.

3 C. Balanis, *Antenna Theory: Analysis and Design*, 2nd ed., John Wiley & Sons, Inc., New York, 1997.

4 A.S. Ramo, J. Whinnery, and T. Van Duzer, *Fields and Waves in Communications Electronics*, 2nd ed., John Wiley & Sons, Inc., New York, 1984.

5 D. Sievenpiper and E. Yablonovitch, *Circuit and method for eliminating surface currents in metal*. US Patent 60/079953, March 30, 1998.

6 L. Brillouin, Wave guides for slow waves. *J. Appl. Phys.*, Vol. 19, 1948, pp. 1023–1041.

7 M.N.M. Tan, M.T. Ali, S. Subahir, T.A. Rahman, and S.K.A. Rahim, Backlobe reduction using mushroom-like EBG structure, in *Proc. of the IEEE Symposium on Wireless Technology and Applications (ISWTA)*, September 23–26, 2012, pp. 206–209.

8 X. Chen, Z.J. Su, L. Li, and C.H. Liang, Radiation pattern improvement in closely-packed array antenna by using mushroom-like EBG structure, in *IET International Radar Conference 2013*, April 14–16, 2013, pp. 1–3.

9 C. Neo and Y. H. Lee, Patch antenna enhancement using a mushroom-like EBG structures, in *2013 IEEE Antennas and Propagation Society International Symposium (APSURSI)*, July 7–13, 2013, pp. 614–615.

10 M. Coulombe, S.F. Koodiani, and C. Caloz, Compact elongated mushroom (EM)-EBG structure for enhancement of patch antenna array performances. *IEEE Trans. Antennas Propag.* Vol. 58, No. 4, 2010, pp. 1076–1086.

11 M.Z. Azad and M. Ali, Novel wideband directional dipole antenna on a mushroom like EBG Structure. *IEEE Trans. Antennas Propag.*, Vol. 56, No. 5, 2008, pp. 1242–1250.

12 B. Archambeault, *PCB Design for Real-World EMI Control*, Kluwer Academic Publishers, Norwell, MA, 2002.

13 R. Senthinathan and J. Price, *Simultaneous Switching Noise of CMOS Device and Circuits*, Kluwer Academic Publishers, Norwell, MA, 1994.

14 V. Ricchiuti, Power supply decoupling on fully populated high-speed digital PCBs. *IEEE Trans. Electromagn. Compat.*, Vol. 43, No. 4, 2001, pp. 671–676.

15 W. Cui, J. Fan, Y. Ren, H. Shi, J.L. Drewniak, and R.E. DuBroff, DC power-bus noise isolation with power-plane segmentation. *IEEE Trans. Electromagn. Compat.*, Vol. 45, No. 2, 2003, pp. 436–443.

16 G.-T. Lei, R.W. Techentin, and B.K. Gilbert, High frequency characterization of power/ground-plane structures. *IEEE Trans. Microw. Theory Tech.*, Vol. 47, 1999, pp. 562–569.

17 S.V. Berghe, F. Olyslager, D. de Zutter, J.D. Moerloose, and W. Temmerman, Study of the ground bounce caused by power plane resonances. *IEEE Trans. Electromagn. Compat.*, Vol. 40, No. 2, 1998, pp. 111–119.

18 J.L. Knighten, B. Archambeault, J. Fan, G. Selli, A. Rajagopal, S. Connor, and J.L. Drewniak, PDN design strategies: IV. Sources of PDN noise, in *IEEE EMC Society Newsletter*, No. 212, 2007, pp. 54–64.

19 J.L. Knighten, B. Archambeault, J. Fan, G. Selli, L. Xue, S. Connor, and J.L. Drewniak, PDN design strategies: III. Planes and materials – are they important factors in power bus design? in *IEEE EMC Society Newsletter*, No. 210, 2006, pp. 58–69.

20 J.L. Knighten, B. Archambeault, J. Fan, G. Selli, L. Xue, S. Connor, and J.L. Drewniak, PDN design strategies: II. Ceramic SMT decoupling capacitors – does location matter? in *IEEE EMC Society Newsletter*, No. 208, 2006, pp. 56–67.

21 J.L. Knighten, B. Archambeault, J. Fan, G. Selli, S. Connor, and J.L. Drewniak, PDN design strategies: I. Ceramic SMT decoupling capacitors – what values should I choose? in *IEEE EMC Society Newsletter*, No. 207, 2005, pp. 46–53.

22 M. Xu, T.H. Hubing, J. Chen, T.P. Van Doren, J.L. Drewniak, and R.E. DuBroff, Power-bus decoupling with embedded capacitance in printed circuit board design. *IEEE Trans. Electromagn. Compat.*, Vol. 45, No. 1, 2003, pp. 22–30.

23 V. Ricchiuti, Power bus signal integrity improvement and EMI mitigation on multilayer high-speed digital PCBs with embedded capacitance. *IEEE Trans. Mob. Comput.*, Vol. 2, No. 4, 2003, pp. 314–321.

24 J. Qin, O.M. Ramahi, and V. Granatstein, Novel planar electromagnetic bandgap structures for wideband noise suppression and EMI reduction in high speed circuits. *IEEE Trans. Electromagn. Compat.*, Vol. 49, No. 3, 2007, pp. 661–669.

25 T.L. Wu, C.C. Wang, Y.H. Lin, T.K. Wang, and G. Chang, A novel power plane with super-wideband elimination of ground bounce noise on high speed circuits. *IEEE Microw. Wirel. Compon. Lett.*, Vol. 15, No. 3, 2005, pp. 174–176.

26 S. Shapharnia and O.M. Ramahi, Electromagnetic interference (EMI) reduction from printed circuit boards (PCB) using electromagnetic bandgap structures. *IEEE Trans. Electromagn. Compat.*, Vol. 46, No. 4, 2004, pp. 580–587.

27 F. de Paulis, M.N. Nisanci, and A. Orlandi, Practical EBG application to multilayer PCB: impact on power integrity. *IEEE Electromagn. Compat. Mag.*, Vol. 1, No. 3, 2012, pp. 60–65.

28 T.L. Wu, Y.H. Lin, T.K. Wang, C.C. Wang, and S.T. Chen, Electromagnetic bandgap power/ground planes for wideband suppression of ground bounce noise and radiated emission in high speed circuits. *IEEE Trans. Microw. Theory Tech.*, Vol. 53, No. 9, 2005, pp. 2935–2942.

29 T. Kamgaing and O.M. Ramahi, A novel power plane with integrated simultaneous switching noise mitigation capability using high impedance surface. *IEEE Microw. Wirel. Compon. Lett.*, Vol. 13, No. 1, 2003, pp. 21–23.

30 R. Abhari and G.V. Eleftheriades, Metallo-dielectric electromagnetic bandgap structures for suppression and isolation of the parallel-plate noise in high-speed circuits. *IEEE Trans. Microw. Theory Tech.*, Vol. 51, No. 6, 2003, pp. 1629–1639.

31 A. Tavallaee, M. Iacobacci, and R. Abhari, A new approach to the design of power distribution networks containing electromagnetic bandgap structures. *Electr. Perform. Elect. Packag.*, 2006, pp. 43–46.

32 T. Kamgaing and O.M. Ramahi, Design and modeling of high impedance electromagnetic surfaces for switching noise suppression in power planes. *IEEE Trans. Electromagn. Compat.*, Vol. 47, No. 3, 2005, pp. 479–489.

33 K.H. Kim and J.E. Shutt-Ainé, Analysis and modeling of hybrid planar-type electromagnetic-bandgap structures and feasibility study on power distribution network applications. *IEEE Trans. Microw. Theory Tech.*, Vol. 56, No. 1, 2008, pp. 178–186.

34 J. Choi, V. Govind, R. Mandrekar, S. Janagama, and M. Swaminathan, Noise reduction and design methodology for the mixed-signal systems with alternating impedance electromagnetic bandgap (AI-EBG) structure, in *IEEE MTT-S International Microwave Symposium Digest.*, Long Beach, CA, June 2005, pp P 645-L 651.

35 T.H. Kim, D. Chung, E. Engin, W. Yun, Y. Toyota, and M. Swaminathan, A novel synthesis method for designing electromagnetic bandgap (EBG) structures in packaged mixed signal systems, in *Proc. of the 56th Electronic Components and Technology Conference*, 2006, pp. 1645–1651.

36 E. Rajo-Iglesias, M. Caiazzo, L. Inclán-Sánchez, and P-.S. Kildal, Comparison of bandgaps of mushroom-type EBG surface and corrugated and strip-type soft surfaces. *IET Microw. Antennas Propag.*, Vol. 1, No. 1, 2007, pp. 184–189.

37 L. Liang, C.H. Liang, L. Chen, and X. Chen, A novel broadband EBG using cascaded mushroom-like structure. *Microw. Opt. Technol. Lett.*, Vol. 50, 2008, pp. 2167–2170.

38 T. Kamgaing and O.M. Ramahi, Multiband electromagnetic-bandgap structures for applications in small form-factor multichip module packages. *IEEE Trans. Microw. Theory Tech.*, Vol. 56, No. 10, 2008, pp. 2293–2300.

39 T.L. Wu, J. Fan, F. de Paulis, C.D. Wang, A. Ciccomancini, and A. Orlandi, Mitigation of noise coupling in multilayer high-speed PCB: state of the art modeling methodology and EBG technology. *IEICE Trans. Commun.*, Vol. E93-B, No. 7, 2010, pp. 1678–1689.

40 A. Ciccomancini Scogna, A. Orlandi, V. Ricchiuti, and T.L. Wu, Impact of photonic crystal power/ground layer density on power integrity performance

of high-speed power buses, in *Proc. of the 2009 IEEE Symposium on Electromagnetic Compatibility*, Austin, TX, August 17–21, 2009.

41   A. Ciccomancini Scogna, T.-L. Wu, and A. Orlandi, Noise coupling mitigation in PWR/GND plane pair by means of photonic crystal fence: sensitivity analysis and design parameters extraction. *IEEE Trans. Adv. Packag.*, Vol. 33, No. 3, 2010, pp. 574–581.

42   A. Ciccomancini Scogna, C.-D. Wang, A. Orlandi, and T.-L. Wu, Parallel-plate noise suppression using a ground surface perturbation lattice (GSPL) structure, in *Proc. of the 2010 Asia-Pacific Symposium on Electromagnetic Compatibility*, Beijing, China, April 12–16, 2010.

43   A. Ciccomancini, T.L. Wu, and A. Orlandi, Power noise suppression in mixed signal circuits using a ground surface perturbation lattice (GSPL), in *Proc. of the DesigCon 2011*, Santa Clara, CA, January 31– February 3, 2011.

44   A. Ciccomancini Scogna, A. Orlandi, T.L. Wu, and T.K. Wang, Analysis and design of GHz power noise isolation using 450 rotated photonic crystal fence, in *Proc. of the 2011 IEEE Symposium on Electromagnetic Compatibility*, Long Beach, CA, August 14–19, 2011.

45   W.-T. Huang, C.-H. Lu, and D.-B. Lin, The optimal number and location of grounded vias to reduce crosstalk. *Prog. Electromagn. Res.*, Vol. 95, 2009, pp. 241–266.

46   B. Wu and L. Tsang, Full-wave modeling of multiple vias using differential signaling and shared antipad in multilayered high speed vertical interconnects. *Prog. Electromagn. Res.*, Vol. 97, 2009, 129–139.

47   F. de Paulis, Y.-J. Zhang, and J. Fan, Signal/power integrity analysis for multilayer printed circuit boards using cascaded *S*-parameters. *IEEE Trans. Electromagn. Compat.*, Vol. 52, No. 4, 2010, pp. 1008–1018.

48   B. Wu and L. Tsang, Full-wave modeling of multiple vias using differential signaling and shared antipad in multilayered high speed vertical interconnects. *Prog. Electromagn. Res.*, Vol. 97, 2009, pp. 129–139.

49   V. Ricchiuti, F. de Paulis, and A. Orlandi, An equivalent circuit model for the identification of the stub resonance due to differential vias on PCB, in *Proc. of the IEEE Workshop on Signal Propagation on Interconnects 2009 (SPI'09)*, Strasbourg, France, May 12–15, 2009.

50   A. Jaze, B. Archambeault, and S. Connor, Differential mode to common mode conversion on differential signal vias due to asymmetric GND via configurations, in *Proc. of the IEEE International Symposium on Electromagnetic Compatibility*, August 5–9, 2013, pp. 735–740.

51   F. de Paulis and A. Orlandi, Accurate and efficient analysis of planar electromagnetic band-gap structures for power bus noise mitigation in the GHz band. *Prog. Electromagn. Res. B*, Vol. 37, 2012, pp. 59–80.

52   F. de Paulis, B. Archambeault, S. Connor, and A. Orlandi, Electromagnetic band gap structure for common mode filtering of high speed differential

signals, in *Proc. of the IEC DesignCon 2011*, Santa Clara, CA, January 31–February 3, 2011.

53 L. Raimondo, F. de Paulis, and A. Orlandi, A simple and efficient design procedure for planar electromagnetic bandgap structures on printed circuit boards. *IEEE Trans. Electromagn. Compat.*, Vol. 53, No. 2, 2011, pp. 482–490.

54 F. de Paulis, L. Raimondo, and A. Orlandi, Impact of shorting vias placement on embedded planar electromagnetic bandgap structures within multilayer printed circuit boards. *IEEE Trans. Microw. Theory Tech.*, Vol. 58, No. 7, 2010, pp. 1867–1876.

55 F. de Paulis, L. Raimondo, and A. Orlandi, IR-DROP analysis and thermal assessment of planar electromagnetic band-gap structures for power integrity applications. *IEEE Trans. Adv. Packag.*, Vol. 33, No. 3, 2010, pp. 617–622.

56 D. Di Febo, M.H. Nisanci, F. de Paulis, and A. Orlandi, Impact of planar electromagnetic band-gap structures on IR-DROP and signal integrity in high speed printed circuit boards, in *Proc. of the 2012 International Symposium on Electromagnetic Compatibility (EMC EUROPE)*, Rome, Italy, September 17–21, 2011.

57 M.N. Nisanci, F. de Paulis, D. Di Febo, and A. Orlandi, Practical EBG application to multilayer PCB: impact on signal integrity. *IEEE Electromagn. Compat. Mag.*, Vol. 2, No. 2, 2013, pp. 82–87.

58 A. Ciccomancini Scogna, A. Orlandi, and V. Ricchiuti, Signal and power integrity performances of striplines in presence of 2D EBG planes, in *Proc. of the 12th IEEE Workshop on Signal Propagation and Interconnects (SPI'08)*, Avignon, France, May 2008.

59 F. de Paulis and A. Orlandi, Signal integrity analysis of single-ended and differential striplines in presence of EBG planar structures. *IEEE Microw. Wirel. Compon. Lett.*, Vol. 19, No. 9, 2009, pp. 554–556.

60 F. de Paulis, A. Orlandi, L. Raimondo, and G. Antonini, Fundamental mechanisms of coupling between planar electromagnetic bandgap structures and interconnects in high-speed digital circuits. Part I—microstrip lines, in *2009 International Symposium on Electromagnetic Compatibility (EMC Europe)*, Athens, Greece, June 11–12, 2009.

61 F. de Paulis, L. Raimondo, and A. Orlandi, Signal integrity analysis of embedded planar EBG structures, in *Proc. of the Asia-Pacific EMC 2010*, Beijing, China, April 12–16, 2010.

62 F. de Paulis, L. Raimondo, S. Connor, B. Archambeault, and A. Orlandi, Design of a common mode filter by using planar electromagnetic bandgap structures. *IEEE Trans. Adv. Packag.*, Vol. 33, No. 4, 2010, 994–1002.

63 F. de Paulis, L. Raimondo, S. Connor, B. Archambeault, and A. Orlandi, Compact configuration for common mode filter design based on

electromagnetic band-gap structures. *IEEE Trans. Electromagn. Compat.*, Vol. 54, No. 3, 2012, pp. 646–654.

64 F. de Paulis, L. Raimondo, D. Di Febo, B. Archambeault, S. Connor, and A. Orlandi, Experimental validation of common-mode filtering performances of planar electromagnetic band-gap structures, in *Proc. of the IEEE International Symposium on Electromagnetic Compatibility*, Fort Lauderdale, FL, July 25–30, 2010.

65 S. Connor, B. Archambeault, and M. Mondal, The impact of common mode currents on signal integrity and EMI in high-speed differential data links, in *Proc. of the IEEE International Symposium on Electromagnetic Compatibility*, August 18–22, 2008, pp. 1–5.

66 F. de Paulis, A. Orlandi, L. Raimondo, B. Archambeault, and S. Connor, Common mode filtering performances of planar EBG structures, in *Proc. of the IEEE International Symposium on Electromagnetic Compatibility*, August 17–21, 2009, pp. 86–90.

67 F. de Paulis, L. Raimondo, D. Di Febo, and A. Orlandi, Routing strategies for improving common mode filter performances in high speed digital differential interconnects, in *Proc. of the 2011 IEEE Workshop on Signal Propagation on Interconnects (SPI'11)*, Naples, Italy, May 8–11, 2011.

68 M.H. Nisanci, F. de Paulis, A. Orlandi, B. Archambeault, and S. Connor, Optimum geometrical parameters for the EBG-based common mode filter design, in *Proc. of the 2012 IEEE Symposium on Electromagnetic Compatibility*, Pittsburgh, PA, August 5–10, 2012.

69 F. de Paulis, M. Cracraft, D. Di Febo, M.H. Nisanci, S. Connor, B. Archambeault, and A. Orlandi, EBG-based common-mode microstrip and stripline filters: experimental investigation of performances and crosstalk. *IEEE Trans. Electromagn. Compat.*, Vol. 57, No. 5, 2015, pp. 996–1004.

70 F. de Paulis, M. Cracraft, C. Olivieri, S. Connor, A. Orlandi, and B. Archambeault, EBG-based common-mode stripline filters: experimental investigation on interlayer crosstalk. *IEEE Trans. Electromagn. Compat.*, Vol. 57, No. 5, 2016, pp. 996–1004

71 F. de Paulis, M.H. Nisanci, D. Di Febo, A. Orlandi, S. Connor, M. Cracraft, and B. Archambeault, Standalone removable EBG-based common mode filter for high speed differential signaling, in *Proc. of the IEEE International Symposium on Electromagnetic Compatibility*, Raleigh, NC, August 3–8, 2014, pp. 244–249.

72 M.A. Varner, F. de Paulis, A. Orlandi, S. Connor, M. Cracraft, B. Archambeault, M.H. Nisanci, and D. Di Febo, Removable EBG-based common-mode filter for high-speed signaling: experimental validation of

prototype design, in *IEEE Trans. Electromagn. Compat.*, Vol. 57, No. 4, 2015, pp. 672–679.

73 C. Kodama, C. O'Daniel, J. Cook, F. de Paulis, M. Cracraft, S. Connor, A. Orlandi, and E. Wheeler, Mitigating the threat of crosstalk and unwanted radiation when using electromagnetic bandgap structures to suppress common mode signal propagation in PCB differential interconnects, in *Proc. of the IEEE International Symposium on Electromagnetic Compatibility*, Dresden, August 16–22, 2015, pp. 622–627.

74 F. de Paulis, B. Archambeault, M.H. Nisanci, S. Connor, and A. Orlandi, Miniaturization of common mode filter based on EBG patch resonance, in *Proc. of the IEC DesignCon 2012*, Santa Clara, CA, January 30–February 2, 2012.

75 W.E. McKinzie, N.D. Nair, B.A. Thrasher, M.A. Smith, E.D. Hughes, and J.M. Parisi, 60 GHz patch antenna in LTCC with an integrated EBG structure for antenna pattern improvements, in *Proc. of the 2014 IEEE International Symposium of Antennas and Propagation Society (APSURSI)*, July 6–11, 2014, pp. 1766–1767.

76 C.-H. Tsai and T.L. Wu, A GHz common-mode filter using negative permittivity metamaterial on low temperature co-fire ceramic (LTCC) substrate, in *Proc. of the 2009 IEEE International Symposium on Electromagnetic Compatibility*, August 17–21, 2009, pp. 91–94.

77 T.L. Wu, C.-H. Tsai, and T. Itoh, A novel wideband common-mode suppression filter for gigahertz differential signals using coupled patterned ground structure, in *IEEE Trans. Microw. Theory Tech.*, Vol. 57, No. 4, 2009, pp. 848–855.

78 Q. Liu, S. Xui, and D. Pommerenke, Narrowband and broadband common mode filter based on a quarter-wavelength resonator for differential signals, *IEEE Trans. Electromagn. Compat.*, Vol. 57, No. 6, 2015, pp. 1740–1743.

79 C. Olivieri, F. de Paulis, A. Orlandi, S. Connor, B. Archambeault, and D.J. Pommerenke, Resonant EBG-based common mode filter for LTCC substrates, in *Proc. of the 2016 IEEE International Symposium on Electromagnetic Compatibility*, Ottawa, ON, July 2016.

80 J. Qin and O.M. Ramahi, Wideband SSN suppression and EMI reduction from printed circuit boards using novel planar electromagnetic bandgap structure, in *Proc. of the 2006 IEEE International Symposium on Electromagnetic Compatibility*, Portland, OR, August 2006.

81 Lee, J., Lee, H., Park, K., Chung, B., Kim, J., and Kim, J., Impact of partial EBG PDN on PI, SI and lumped model-based correlation, in *Proc. of the 2008 Asia-Pacific Symposium on Electromagnetic Compatibility and 19th International Zurich Symposium on Electromagnetic Compatibility*, Singapore, May 2008.

82 J.-F. Kiang, *Novel Technologies for Microwave and Millimeter: Wave Applications,*, Springer, New York, 2008, Chapter 12.

83 F. Yang, *Electromagnetic Band Gap Structures in Antenna Engineering*, 1st ed., Cambridge University Press, 2009.

84 Computer Simulation Technology, CST Studio Suite, 2015. Available at www.cst.com.

85 Keysight Technologies, Advanced Design System, 2014. Available at www.keysight.com

2

Planar Ebgs: Fundamentals and Design

Modern digital electronic systems are subject to ever-increasing data rates and associated lower voltage swings, thus leading to a reduced noise margin for ensuring the correct signal propagation. Therefore, the noise mitigation is becoming a must-to-do along the whole design procedure: from the electrical schematic to the printed circuit board (PCB) layout. The integration of numerous functionalities within the same system (at PCB level as well as at the package level) requires high attention for ensuring the required level of noise isolation. Several aspects/noise sources need to be taken into account during the PCB design, such as the well-known simultaneous switching noise (SSN) that can propagate across the PCB through the cavities made by power/ground-reference planes [1] and discontinuities along the high-speed interconnects, such as vias, and imbalances in differential traces [2–10]. Moreover, mixed-signal systems require isolation of the analog circuitry from the digital section to decouple the current return paths in order to avoid spurious harmonic signals to possibly affect the radio frequency (RF) functionalities [11–13]. One solution for reducing the noise propagation within the PCB power planes relies on the use of electromagnetic bandgap (EBG) structures, first introduced for isolating adjacent antennas [14–20]. They have also been considered as a simple and easy-to-design approach to be implemented in PCBs for SSN rejection in the gigahertz range, where usual techniques for power bus decoupling, that is, bypass capacitors, are not effective [21]. Various strategies/layout geometries have been adopted for designing the EBGs to reduce noise propagation within the power planes in PCB and packages [22–29].

The planar EBG is made by a sequence of patches connected by narrow bridges, thus altering the ideal solid plane pair geometry usually employed for power delivery purposes and signal/power return. The patterned plane, together with an adjacent solid plane, builds a cavity with a frequency response characterized by a bandgap. Several methods (based on equivalent circuits and/or on the dispersion diagrams) have been proposed, for analyzing the behavior of the EBG structures [30–33]. This chapter focuses on the comprehensive

Electromagnetic Bandgap (EBG) Structures: Common Mode Filters for High-Speed Digital Systems,
First Edition. Antonio Orlandi, Bruce Archambeault, Francesco De Paulis, and Samuel Connor.
© 2017 by The Institute of Electrical and Electronics Engineers, Inc. Published 2017 by John Wiley & Sons, Inc.

electromagnetic characterization of planar EBGs, concentrating on the simple yet effective EBG made by a sequence of square patches and straight bridges. This analysis turns out to be necessary for a deep comprehension of the physical mechanisms guiding the use of planar EBGs as common mode filters, as accomplished in Refs [34,35].

2.1 Fundamental Behavior of Planar EBG

The planar EBG structure alters the typical geometry of two adjacent solid planes that are commonly used for power delivery purposes in multilayer PCB. The resonant behavior of a cavity made by these two solid planes can be modified if one of the two planes is etched accordingly by obtaining a sequence of square patches connected by narrow bridges.

The resonant behavior of the typical cavity made by two adjacent solid power planes in multilayer PCB is modeled as a cavity having perfect electric conductor (PEC) boundary conditions at the top and bottom walls (power planes) and perfect magnetic conductor (PMC) boundary conditions at the side walls [4,5], as in Figure 2.1.

The common dimensions in multilayer PCB such as the thin dielectric between the two planes ($d < <A$, $d < <B$, $d < <\lambda$, where λ is the wavelength associated with the frequency of interest) on the order of few mils (hundreds of micrometers) leads to simplify the solution of the Helmholtz equations, and thus of the dispersion relation, leading to the expression in (2.1) [36] for the allowed frequencies associated with the resonant TM modes inside the cavity:

$$f_{TM_z,mn} = \frac{c}{2\pi\sqrt{\varepsilon_r}}\sqrt{\left(\frac{m\pi}{A}\right)^2 + \left(\frac{n\pi}{B}\right)^2} \tag{2.1}$$

where A and B are the dimensions of the cavity along the x and y dimensions, ε_r is the electric permittivity of the dielectric, and c is the speed of light. The modes are identified as TM_z since the thin cavity does not allow the propagation of

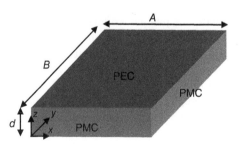

Figure 2.1 Ideal cavity with top and bottom PEC walls and side PMC walls that models the cavity created between power planes in multilayer PCBs.

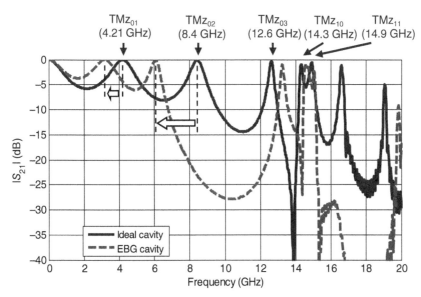

Figure 2.2 Transfer function of the cavity in Figure 2.1, and of the EBG cavity with the top plane etched as in Figure 2.3. (Taken from Ref. [34]. Reproduced with permission of The Electromagnetics Academy.)

modes with z variation, thus the electric and magnetic fields are constant along the z direction. The typical behavior of a cavity is reported in Figure 2.2, solid curve, obtained by CST Studio [37], a three-dimensional electromagnetic (EM) solver simulating a cavity with the following dimensions: $A = 5$ mm, $B = 17$ mm, dielectric thickness $d = 0.4$ mm, and $\varepsilon_r = 4.4$. The result in the figure shows the transfer function in terms of S-parameters, the insertion loss $|S_{21}|$, between two ports inside the cavity located at port 1 (2 mm; 2 mm) and port 2 (3 mm; 15 mm). The port is defined as a vertical excitation from the bottom to the top PEC walls. Figure 2.2 also includes the first five resonant mode frequencies calculated using (2.1), perfectly corresponding to the peaks in the $|S_{21}|$ curve.

Moving from the ideal case in Figure 2.1 to the cavity with the etched top plane, as the one in Figure 2.3, the electromagnetic field inside the cavity at the

Figure 2.3 Etched top plane made by $N = 3$ square patches ($a = b = 5$ mm) connected by two narrow bridges ($w = 0.5$ mm, $g = 1$ mm) with outline as the cavity in Figure 2.1 with $A = 5$ mm, $B = 17$ mm. (Taken from Ref. [34]. Reproduced with permission of The Electromagnetics Academy.)

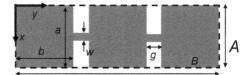

resonant frequencies is affected by the altered geometry since the conduction current associated with the field is forced to flow through the narrow bridges. The simulation ports in the model in Figure 2.3 are placed in the lateral patches. This effect impacts on the resonant frequencies that are shifted to low frequencies as we can see from the dashed curve results in Figure 2.2. The etched plane in Figure 2.3 is made by $N = 3$ square patches along the y-direction.

Figure 2.4a and b shows the current density $|J_y|$ on the bottom cavity plane at the first resonant mode, TM_{z01}, that occurs at 4.21 GHz for the ideal cavity and at 3.2 GHz for the EBG cavity. The main difference between the two charts is that the current density has a maximum at the bridges, much larger than the maximum for the ideal cavity (210 versus 105 A/m), even though the trend typical of the TM_{z01} mode is satisfied with a maximum at the cavity center and minima at the edges ($y = 0$ and $y = 17$ mm). Similar conclusions can be carried out looking at the electric field $|E_z|$ inside the cavities given in Figure 2.4c and d. Here, the field maxima are at the edges with a minimum at the cavity center, as expected.

These differences in the electromagnetic behavior explain the differences between the curves in Figure 2.2. The peaks are moved toward lower frequencies for the patterned plane cavity. A circuit interpretation of this behavior can be given introducing the concept of an additional inductance associated with the bridges compared to the inductance of the solid cavity [38]. The shift impacts only some of the resonant modes, up to the mode $TM_{z0,N-1}$, where N is the number of patches along the y-direction. Beyond this point, there is the mode with index equal to N; this mode can be associated either with the whole patterned cavity or as the first resonance of the single patch cavity, that is, the small cavity made by each square patch and the solid plane underneath (the mode $TM_{z0,N}$ of the whole cavity corresponds to the mode TM_{z10} (TM_{z01}) of the single patch cavity). Therefore, a bandgap is generated between the $TM_{z0,N-1}$ mode and the first resonance mode TM_{z10} of the single patch cavity. The mode $TM_{z0,N-1}$ of the EBG cavity identifies the lower limit of the bandgap, f_{Low}, and the first resonant mode of the single patch identifies the upper limit of the bandgap, f_{High}. The accurate quantification of these limits represents the key point for efficiently designing the EBG structure as a technique for noise reduction in PCB for high-speed digital and mixed systems.

2.2 Identification of the Bandgap Limits: f_{High}

The upper limit of the bandgap, f_{High}, is associated with the single patch cavity, as the first resonant mode TM_{z10} [37]. It can be easily identified by substituting the proper parameters into (2.1) thus obtaining (2.2):

$$f_{High} = f_{TM_z,10} = \frac{c}{2a\sqrt{\varepsilon_r}} \tag{2.2}$$

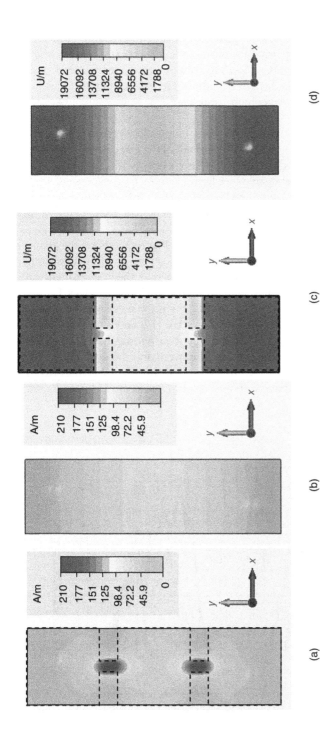

Figure 2.4 Distribution of the current density $|J_y|$ for (a) the EBG cavity and (b) the ideal cavity. Distribution of the electric field $|E_z|$ for (c) the EBG cavity and (d) the ideal cavity. The components $|J_y|$ and $|E_z|$ are the only nonzero, as expected for the TM_{201} mode.

where a is the size of the square patch. This equation, widely used for the identification of the bandgap upper limit, does not describe exactly the EM behavior of an EBG structure, even though it can be considered a good approximation. A more quantitative study is carried out herein focusing on the relationship between the geometry and the first resonant mode beyond the bandgap.

A deeper investigation into the physics of the first resonances after the bandgap leads us to take into account the effect of the bridges and their impact on the f_{High}. The TM_{z10} mode of the single patch cavity is affected by the bridges that lead to a shift of the bandgap upper limit from the f_{High} value given in (2.2). Some additional models are simulated based on the $1 \times N$ array of patches with $N = 5$ for quantifying the impact of the bridges. The patch size is the same as from the geometry given in Figure 2.3 ($a = 5$ mm), whereas the bridge dimensions w and g are varied for a parametric analysis. Figure 2.5 shows the results of four models with variation of the bridge length g, keeping its width w constant to 0.5 mm. The f_{High} value from (2.2) is 14.3 GHz; however, the first resonance after the bandgap is at lower frequency, and it is smaller for longer bridges. Figure 2.6 reports the pattern of the electric field $|E_z|$ at the resonances occurring at 13.4, 13, 12.3, and 11.8 GHz for the cases with $g = 0.5$, 1, 2, and 3 mm, respectively. The pattern is very similar for the four figures, noting that the amplitude increases for the shorter bridge models, thus verifying the correctness of the nature of the resonance related to the TM_{z10} of the single patch cavity. The quantification of the frequency difference is offered in Figure 2.7, which

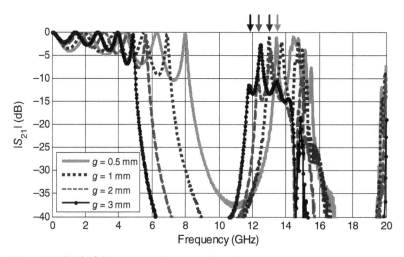

Figure 2.5 $|S_{21}|$ of the 5×1 patch array with $a = b = 5$ mm, $w = 0.5$ mm, $g = 0.5$, 1, 2, 3 mm. (Taken from Ref. [34]. Reproduced with permission of The Electromagnetics Academy.)

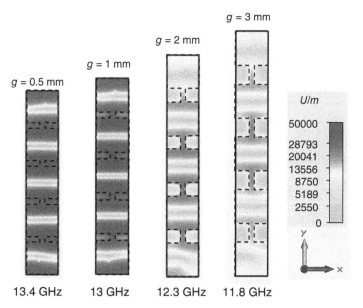

Figure 2.6 $|E_z|$ pattern at the first resonance after the bandgap, occurring at 13.4, 13, 12.3, and 11.8 GHz for the cases with $g = 0.5$, 1, 2, and 3 mm, respectively. (Taken from Ref. [34]. Reproduced with permission of The Electromagnetics Academy.)

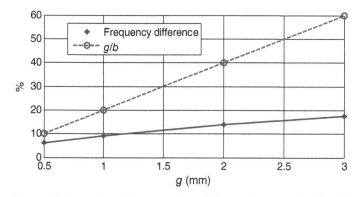

Figure 2.7 Evaluation of the percentage difference between the ideal TM_{z10} resonant frequency 14.3 GHz of the single patch cavity, and the first resonance after the bandgap for the four considered models (solid curve). Percentage ratio between the bridge length g and the patch size $a = b$ (dashed curve). (Taken from Ref. [34]. Reproduced with permission of The Electromagnetics Academy.)

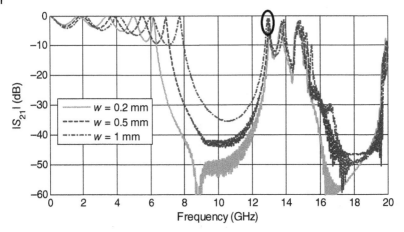

Figure 2.8 $|S_{21}|$ of the 1×5 patch array with $a = b = 5$ mm, $g = 1$ mm, $w = 0.2, 0.5, 1$ mm. (Taken from Ref. [34]. Reproduced with permission of The Electromagnetics Academy.)

summarizes the resonance values and the percentage difference from the ideal 14.3 GHz value.

The difference in the frequency value can be kept within a certain limit (i.e., below 10%) by designing a bridge that is no longer than the 20% of the patch size.

An additional test is carried out keeping constant the bridge length at $g = 1$ mm, and varying the bridge width $w = 0.2, 0.5$, and 1 mm. The results are shown in Figure 2.8. The bridge width does not impact the f_{High} since all the three models provide an f_{High} around 13 GHz. The inductance associated with the three bridges still changes, increasing for narrower bridges, thus leading to larger shift toward lower frequency of the first $N - 1$ resonant modes. An accurate design of an EBG in terms of bandgap upper limit can be done referring to (2.2) and trying to keep the bridge length as small as possible, to keep the f_{High} closer to the TM_{z10} mode of the single patch cavity.

2.3 Identification of the Bandgap Limits: f_{Low}

The calculation of the lower limit of the bandgap, f_{Low}, is more difficult and requires a deeper study. However, based on the concept of the excess of inductance associated with the narrow bridges, in Ref. [38] the value of f_{Low} has been quantified through an expression similar to (2.2).

The main ideas in Ref. [38] are briefly recalled for clarity as follows. The concept of the bridge inductance as a factor contributing to increase the overall cavity inductance is validated in Ref. [38]; some results are provided for different bridge width, demonstrating that a larger bridge has associated a lower inductance value, thus achieving smaller frequency shift of the first $N-1$ resonant modes. Along with this behavior, the measured structures show that the upper limit of the bandgap remains unchanged [38], as demonstrated also in Section 2.2; therefore, the excess of inductance impacts the bandgap width.

The concept that only the first $N-1$ resonant modes are shifted down in frequency is confirmed by the results shown previously, that is, in Figure 2.2 where the first two modes of the 1×3 EBG geometry are moved from 4.21 to 3.2 GHz (TM_{z01}) and from 8.4 to 6.1 GHz (TM_{z02}), leaving a bandgap from 6.1 to 13.2 GHz. Similar trend is found looking at the results in Figure 2.5 and Figure 2.8 providing the results for the 1×5 EBG, where the bandgap is obtained after the shift of the first four modes. This concept can be extended when considering a two-dimensional geometry, that is, an $M \times N$ matrix of patches. Therefore, the modes that will be affected by the shift toward lower frequency are those with index less than M and N, thus the last mode before the bandgap can be identified as the $\mathrm{TM}_{zM-1,N-1}$. A simple model made by a 2×3 patch matrix is simulated and it is shown in Figure 2.9; the results in terms of $|S_{21}|$ are provided in Figure 2.10 (dashed curve). The geometry is based on the patch and bridge dimensions as in Figure 2.3, with $a = b = 5$ mm, $g = 1$ mm, and $w = 0.5$ mm; the stack-up parameters are the dielectric thickness $d = 0.4$ mm, metal thickness $t = 0.017$ mm, and dielectric permittivity $\varepsilon_r = 4.4$.

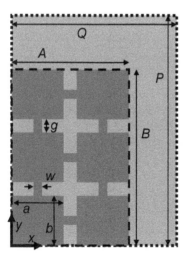

Figure 2.9 Top view of the model for the 2×3 EBG matrix (blue color) with $a = b = 5$ mm, $g = 1$ mm, and $w = 0.5$ mm, Port 1 at (2 mm, 2 mm), Port 2 at (9 mm, 15 mm). Its equivalent solid plane counterpart (green color), Port 1 at (2 mm, 2 mm), Port 2 at (11.8 mm, 20.6 mm). (Taken from Ref. [34]. Reproduced with permission of The Electromagnetics Academy.)

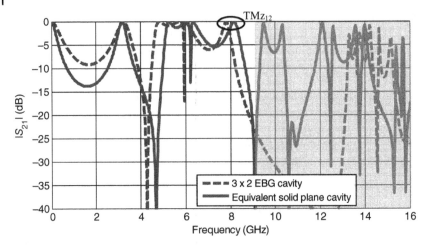

Figure 2.10 $|S_{21}|$ of the 2×3 EBG matrix with $a = b = 5$ mm, $g = 1$ mm, $w = 0.5$ mm. (Taken from Ref. [34]. Reproduced with permission of The Electromagnetics Academy.)

The first resonant modes that are shifted, up to the $\text{TM}_{zM-1,N-1}$ mode, could be considered as associated with a solid plane cavity with dimensions larger than the overall size of the patterned cavity, such as length $P > B$ and $Q > A$. If the P and Q values can be determined, then the f_{Low} can be seen as the mode $\text{TM}_{M-1,N-1}$ of the equivalent enlarged geometry, as in (2.3).

$$f_{TM_z,M-1,N-1} = \frac{c}{2\sqrt{\varepsilon_r}} \sqrt{\left(\frac{M-1}{Q}\right)^2 + \left(\frac{N-1}{P}\right)^2} \qquad (2.3)$$

The idea of the equivalent enlarged cavity based on the concept of equivalent total inductance is described analytically in (2.4), assuming one-dimensional array of patches along both the x- and y-directions [38]. The total inductance along each direction is computed as sum of the inductances of the M (N) patches and the $M - 1$ ($N - 1$) bridges, and it is set equal to the inductance of a parallel plane transmission line (PPTL) of length Q (P) and width b (a), as defined in (2.5):

$$L_{tot,X} = \mu_0 d \frac{Q}{b} = M L_{patch,X} + (M-1) L_{bridge} \qquad (2.4a)$$

$$L_{tot,Y} = \mu_0 d \frac{P}{a} = N L_{patch,Y} + (N-1) L_{bridge} \qquad (2.4b)$$

$$L_{PPTL} = \mu_0 d \frac{Length_{PPTL}}{Width_{PPTL}} \tag{2.5}$$

$$L_{PPTL} = \mu_0 d \tag{2.6}$$

where d is the dielectric thickness and μ_0 is the magnetic permeability in free space. Equation (2.5) is also employed for calculating the L_{patch} in (2.4); for square patches, thus for a patch having equal values of length and width, (2.5) reduces simply to (2.6).

The dimensions P and Q of the equivalent solid plane cavity can be easily derived from (2.4), as in (2.7):

$$Q = \frac{b}{\mu_0 d} \left(N L_{patch,X} + (N-1) L_{bridge} \right) \tag{2.7a}$$

$$P = \frac{a}{\mu_0 d} \left(M L_{patch,Y} + (M-1) L_{bridge} \right) \tag{2.7b}$$

The bridge can be approximated as microstrip transmission line [36], whose inductance can be computed as in (2.8).

$$L_{bridge} = \begin{cases} l \cdot \dfrac{60}{c_0} \ln \left(\dfrac{8d}{w} + \dfrac{w}{4d} \right), & \dfrac{w}{d} \le 1 \\[4mm] l \cdot \dfrac{120\pi}{c_0} \left[\dfrac{w}{d} + 1.393 + 0.667 \ln \left(\dfrac{w}{d} + 1.444 \right) \right]^{-1}, & \dfrac{w}{d} \ge 1 \end{cases} \tag{2.8}$$

where c_0 is the speed of light in free space. The lower limit of the bandgap, $f_{Low} = f_{TMz,M-1,N-1}$ can be computed as in (2.3) by using (2.7). This procedure is applied to the 2×3 EBG structure in Figure 2.9 obtaining the equivalent inductances $L_{tot,X} = 1.38\,nH$ and $L_{tot,Y} = 2.27\,nH$ that result, after applying (2.7), in the following equivalent solid plane dimensions $P = 13.8\,mm$ and $Q = 22.6\,mm$. The equivalent geometry is then simulated obtaining the solid curve in Figure 2.10. The first modes, up to the one with index $m = M - 1 = 1$ and $n = N - 1 = 2$, occur at frequencies close to those of the patterned 3×2 EBG cavity. The percentage error between each one of the first five modes are 2.1% (TM_{z01}), 8.9% (TM_{z10}), 4.5% (TM_{z11}), 5.7% (TM_{z02}), and 4.6% (TM_{z12}). The error is always below 10% and the last mode TM_{z12} that is related to the identification of f_{Low} has an error less than 5%.

2.4 Characterization of f_{Low} for Different Patch Matrix Configurations

A complete analysis is carried out in this section studying the lower limit of the bandgap, f_{Low}, for different patch configurations.

The simplest geometry is the 1D sequence of patches, thus achieving a $M \times 1$ ($1 \times N$) array. However, a more general EBG configuration is based on the 2D patch matrix, as the 2×3 EBG structure in Figure 2.9. This section aims at providing a complete characterization of EBG configurations for geometries made by many patches, thus theoretically for M and/or $N \rightarrow \infty$. This approach leads to interesting results useful for a simple and efficient selection of the geometry parameters, for achieving the desired bandgap limits together with the allowed number of patches to fit the available layout area.

The simplest geometry, as already introduced in Section 2.1, is the 1D array of patches. The next analytical developments are based on the $M \times 1$ array, thus on the patches placed sequentially along the x-direction. Equation (2.3) can be combined to (2.7a) to obtain (2.9), and then the limit for $M \rightarrow \infty$ can be evaluated as in (2.10):

$$f_{Low} = f_{TM_{M-1,0}} = \frac{c_0}{2\sqrt{\varepsilon_r}} \frac{M-1}{Q} = \frac{c_0}{2\sqrt{\varepsilon_r}} \frac{M-1}{M \cdot L_{patch} + (M-1) \cdot L_{bridge}} \frac{\mu_0 d}{b}$$

(2.9)

$$f_{Limit} \triangleq \lim_{M \rightarrow \infty} f_{Low} = \frac{c_0 \mu_0 d}{2b\sqrt{\varepsilon_r}} \lim_{M \rightarrow \infty} \frac{M-1}{M \cdot L_{patch} + (M-1) \cdot L_{bridge}}$$
$$= \frac{c_0 \mu_0 d}{2b\sqrt{\varepsilon_r}} \frac{1}{L_{patch} + L_{bridge}}$$

(2.10)

The f_{Limit} in (2.10) does not depend on M, but it represents the upper limit for f_{Low} for an increasing number of patches. More details and a quantitative explanation on the error offered by (2.9) and (2.10) can be found in Ref. [34]. Moreover, in Ref. [34], the evaluation of f_{Low} and f_{Limit} for the general case of $M \times N$ matrix-based EBG is provided as well as the validation of (2.9) and (2.10) through tens of full-wave simulations.

2.5 Experimental Validation

Three test boards were built to validate the proposed analytical approach. The stack-up geometry is characterized by dielectric thickness $d = 0.508$ mm, $\varepsilon_r = 3$, and $tg\delta = 0.0015$. The bridge dimensions are maintained constant for the three

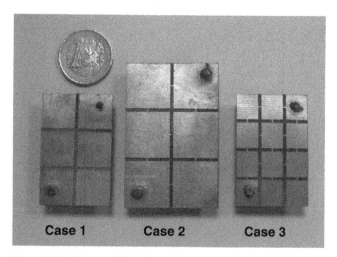

Case 1   Case 2   Case 3

Figure 2.11 Pictures of the three EBG boards. (Taken from Ref. [34]. Reproduced with permission of The Electromagnetics Academy.)

cases, $g = 1.3$ mm, $w = 0.4$ mm. The differences among the three models are as follows:

Case 1: $a = b = 13.7$ mm. $M = 3$, $N = 2$.
Case 2: $a = b = 18$ mm. $M = 3$, $N = 2$.
Case 3: $a = 9.95$ mm, $b = 8.7$ mm. $M = 4$, $N = 3$.

Figure 2.11 shows the three test boards. The solder balls identify the inner pin of the SMA connector that are mounted on the back solid layer. The boards are measured with a 50 MHz to 9 GHz VNA (Anritsu MS4624B). The measurement results are given in Figure 2.12. The Case 1 test board is modeled and the simulation results are included in Figure 2.12 (dotted line). The curves related to the measured and simulated Case 1 test board agree well to each other. The f_{High} for the Case 3 is not very clear since the SMA connector is placed close to the patch center, and thus the TM_{10} mode of the single patch cavity is weakly excited; the f_{High} occurs at 8.31 GHz, with an amplitude below −60 dB. The bandgap limits are extracted from the measured data and they are included in Table 2.1. These results are compared with those computed by applying (2.3)–(2.8) for the f_{Low}, and by applying (2.2) for the f_{High}. The errors for the Case 1 and Case 2 ($M = 3$, $N = 2$) are larger than the Case 3 ($M = 4$, $N = 3$). The error is larger for Case 2 since it consists of a larger patch (17.5% error, $a = 18$ mm) with respect to Case 1 (13.6% error, $a = 13.7$ mm), as could be expected from the error quantification given in Ref. [34]. The error associated with the analytical evaluation of the f_{High} is very small, below 5%. The error is proportional to the ratio between the bridge length g and patch size a (b), as stated in Figure 2.7. It

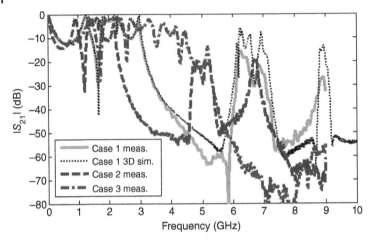

Figure 2.12 Measured $|S_{21}|$ for the three EBG boards. (Taken from Ref. [34]. Reproduced with permission of The Electromagnetics Academy.)

Table 2.1 Bandgap limits f_{Low} and f_{High} extracted from the measured data and computed through the proposed approach.

Test case	Data type	f_{Low} (GHz)	Error (f_{Low})	f_{High} (GHz)	Error (f_{High})
1. 3×2	Analytical	3.35	13.6%	6.32	1.12%
($a = 13.7$ mm)	Measurements	2.95		6.25	
2. 3×2 ($a = 18$ mm)	Analytical	2.55	17.5%	4.8	1.7%
	Measurements	2.17		4.72	
3. 4×3	Analytical	5.56	6.9%	8.7	4.63%
	Measurements	5.2		8.31	

Error evaluation between the two data sets.

goes from 0.84% for Case 2 ($g/b = 7.2\%$) to 1.7% for Case 1 ($g/b = 9.5\%$), to 4.63% for Case 3 ($g/b = 13\%$).

A main drawback when implementing an EBG structure covering partially or fully a PCB plane intended for power delivery is its impact on the plane resistance. The narrow bridges, while performing well at high frequency, represent a bottleneck in the DC resistance path from the voltage regulator module to the ICs requiring the specific power supply voltage associated with the plane. Moreover, the resistance increase leads to thermal issues that need to be carefully addressed taking into account the estimated current supplied by the voltage rail. A comprehensive analysis of both the IR drop and the quantification of thermal aspects related to the use of planar EBGs can be found in Ref. [35].

The EBG structures discussed herein deal with patterned planes laid out on the outer stack-up layers of a multilayer PCB. The inclusion of an EBG area in a PCB inner layer involves additional aspects that need to be addressed before the EBG can still be effectively employed. The problems related to this type of layout are introduced in Ref. [39] and are discussed in the next section.

2.6 Embedded Planar EBG

Few works mention the case where the EBG layer is laid out in an inner layer of a typical multilayer PCB: among them particular attention must be given to Refs [40–42]. This configuration, in general, has the patterned plane sandwiched between two solid layers; thus, it is referred to as an embedded EBG structure. The possibility of overcoming the limitation of the exclusive use of the top layer position, along with the advantages of planar EBG structures employing only a single layer, will increase the flexibility of adapting such a structure as a noise filter in multilayer PCB.

In Ref. [40], the performance of an embedded EBG is studied with respect to the presence of vias connecting the top and bottom solid planes to one another. In this study, the vias are clustered around excitation ports. In the last section of Ref. [41], the case of an embedded EBG is briefly considered. The filtering performance is maintained by employing shorting vias connecting the upper and lower solid planes. Also, in this configuration, the shorting vias surround the test ports. However, a demonstration on how this configuration performs was not provided, and the main physical phenomena that are involved were not investigated. Reference [42] offers a multiconductor transmission line-based analysis of an embedded (named "shielded") EBG. In this paragraph, it is demonstrated that the presence of a third solid plane, laid out on top of the patterned one and that is not electrically shorted to the bottom solid plane, significantly reduces the bandgap.

The physical mechanisms governing the electromagnetic behavior of embedded planar EBG when the patterned plane is laid out on an inner stack-up layer, typically having one solid plane above and below it, are systematically investigated. The study carried out considers the different positioning of the shorting vias between the two solid planes. The proposed layout technique ensures that an embedded EBG structure behaves as a typical planar structure. This allows the use of the same design guidelines as proposed in the previous paragraphs, extending the different design strategies proposed in the past to the more general case of embedded EBG structures.

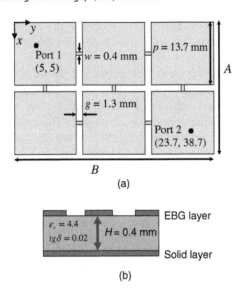

Figure 2.13 (a) Top view of the patterned plane of a planar EBG: $A = 43.7$ mm, $B = 28.7$ mm and the ports' positions are in mm. (b) Board stack-up. (c) Regions of the frequency spectrum of $|S_{21}|$. (Taken from Ref. [39]. Reproduced with permission of IEEE.)

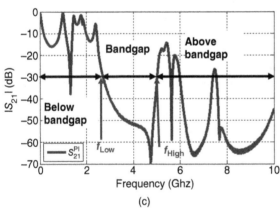

2.6.1 Preliminary Considerations on Embedded EBG

The simple geometry given in Figure 2.13 is laid out on an inner layer between two solid planes, and the structure in Figure 2.14 is obtained. This configuration consists of the patterned plane and the top and bottom solid planes, generating two subcavities. The dimensions of the patterned EBG plane are those given in Figure 2.13. The thickness of the dielectric above and below this layer is 0.4 mm and its properties are $\varepsilon_r = 4.4$ and $tan\delta = 0.02$.

A simple two solid plane configuration is also considered by removing the patterned layer and defining the test port 1 and the test port 2 at the location

Figure 2.14 Embedded EBG layer in between two solid planes. (Taken from Ref. [39]. Reproduced with permission of IEEE.)

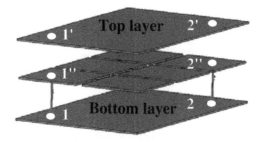

1–1′ and 2–2′, respectively, as shown in Figure 2.14. Figure 2.15 illustrates $|S_{21}|$ associated with the two solid plane cavities. This result is then compared with the transfer function of the embedded EBG structure in Figure 2.14, from port 1 (at 1–1″) to port 2 (at 2–2″). The bandgap between 2.7 and 5 GHz, as shown in Figure 2.13c, disappears and is replaced by distributed resonances (solid black curve). These resonances at 2.55, 2.97, 3.31, and 4.11 GHz correspond to the TM_z modes (TM_{01}, TM_{11}, TM_{20}, and TM_{21}, respectively) of the solid plane cavity. Therefore, the EBG layer between the two solid planes becomes completely inefficient within this band, thus validating the results in Ref. [42].

This behavior is also proven by looking, in Figure 2.16, at the spatial distribution of $|E_z|$, the vertical component of the electric field, inside the embedded EBG cavities at the frequency of 3.31 GHz (TM_{11} mode). This frequency belongs to the bandgap (see Figure 2.13c) of the corresponding

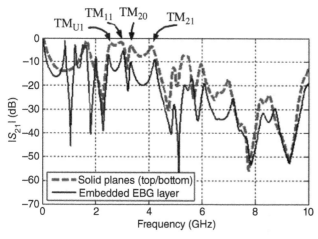

Figure 2.15 Comparison of $|S_{21}|$ of the solid plane cavity and the EBG layer embedded structure within the two solid planes. (Taken from Ref. [39]. Reproduced with permission of IEEE.)

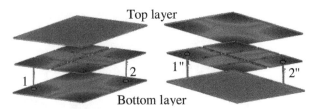

Top layer

Bottom layer

Figure 2.16 Spatial distribution of $|E_z|$ at 3.31 GHz (TM$_{11}$) exciting the lower subcavity. (a) *Top perspective view:* Field pattern at the EBG and at the bottom layers. (b) *Bottom perspective view:* Field pattern at the top and at the EBG layers. (Taken from Ref. [39]. Reproduced with permission of IEEE.)

planar, not embedded, EBG. Two conclusions can be drawn: (i) The propagation of the electric field at a certain frequency (i.e., the frequency of the TM$_{11}$ mode shown in Figure 2.16) is inhibited in the planar EBG and it is allowed (as a TM$_{11}$ mode) in the embedded case due to the effect of the additional solid third plane making a new solid plane cavity. (ii) The same field pattern (associated with the TM$_{11}$ mode) is visible on each of the three planes (bottom, patterned, top) because the presence of the middle EBG plane would not have any effect. The resonant behavior of the cavity made by the top and the bottom solid planes dominates; these resonances must then be inhibited in order to restore the desired bandgap, as for the case of conventional planar EBG structure. This effect is achieved by shorting the two solid planes by an array of vias, inhibiting the low-frequency resonances of the two solid plane cavity. This implies that the two solid planes should have been assigned the same voltage reference in the design of the stack-up. A circular gap is etched at the EBG layer around the via barrel to isolate the via from the patterned plane.

Thirteen shorting vias are added to the model in Figure 2.14, following a regular pattern; the top view of the new model is shown in Figure 2.17. Their mutual distances are evaluated as explained in the following. The via radius is 25.4 μm and the etched gap radius at the EBG layer is 50 μm. The via at

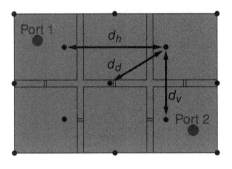

Figure 2.17 Top view of the embedded EBG layer with the location of the 13 shorting vias. (Taken from Ref. [39]. Reproduced with permission of IEEE.)

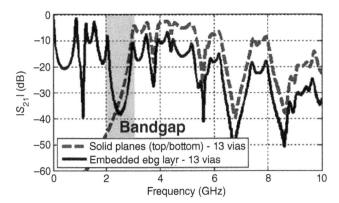

Figure 2.18 $|S_{21}|$ of the solid planes (continuous) and $|S_{21}|$ with the embedded EBG layer (dashed) with 13 shorting vias. (Taken from Ref. [39]. Reproduced with permission of IEEE.)

the plane center falls on the bridge and thus it is shifted 1 mm to the left. $|S_{21}|$ of the two solid plane cavity, with the 13 shorting vias (without the patterned layer), and $|S_{21}|$ of the geometry including the patterned layer are compared in Figure 2.18. The dashed curve starts at DC from $-\infty$, as the two planes are shorted; it then increases monotonically and shows a first resonant peak at 3.05 GHz. The continuous curve shows the resonances between DC and 2.0 GHz (the below bandgap region), related to the bottom subcavity. A narrow but evident bandgap between 2.0 and 3.05 GHz is then generated. The resonances of the two solid planes' cavity control the response of the embedded EBG from 3.05 GHz (the above bandgap region). In this case, the embedded EBG shows a bandwidth $BW = 3.05 - 2 = 1.05$ GHz, smaller than $BW = 2.3$ GHz of the planar EBG in Figure 2.13c. The f_{high} of the embedded EBG structure is related to the first (lowest) resonance that occurs in the two solid plane cavity, including the 13 shorting vias. Figure 2.19 shows the $|E_z|$ spatial distribution at the two cut planes in the middle of the upper and lower subcavities. The field amplitude assumes a maximum value between two adjacent shorting vias, whereas the minimum is found at the via locations due to the high conductivity (of copper) of the via barrel. This spatial distribution can be approximated as a "$\lambda/2$" resonance (λ being the wavelength) occurring between two vias, as schematically illustrated in Figure 2.20, having associated a resonant frequency given by

$$f = \frac{1}{2d\sqrt{\mu\varepsilon}} \tag{2.11}$$

in which μ and ε are the magnetic permeability and electric permittivity of the dielectric and d is the distance between two vias. The distance between two adjacent vias is not the same for each via pair, thus the values of d_v, d_h, and d_d,

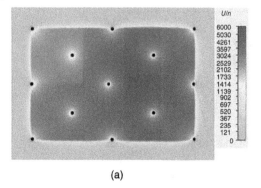

(a)

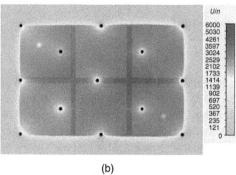

(b)

Figure 2.19 Top view of the spatial distribution of $|E_z|$ in the (a) upper and (b) lower subcavities at 3.05 GHz. (Taken from Ref. [39]. Reproduced with permission of IEEE.)

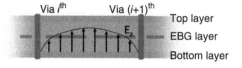

Figure 2.20 Lateral view of the spatial distribution of $|E_z|$ between two shorting vias. (Taken from Ref. [39]. Reproduced with permission of IEEE.)

indicating the vertical, horizontal, and diagonal distances, respectively, are introduced.

The configuration considered in Figure 2.17 is characterized by the following parameters: $d_v = 14.38$ mm, $d_h = 21.85$ mm, and $d_d = 14.8$ mm. The largest value among these three distances is d_h; thus, this distance is considered to compute the first (lowest) resonance of the solid plane cavity. The largest possible distance d between the stitching vias can be computed from (2.11), as in (2.12), given the values of f_{High}:

$$d = \frac{1}{2f_{High}\sqrt{\mu\varepsilon}} \tag{2.12}$$

By looking at the continuous curve in Figure 2.18, substituting $f = 3.05$ GHz in (2.12), one obtains $d = 23.8$ mm. This value is very close to the designed $d_h = 21.85$ mm.

In order to evaluate the largest distance between the shorting vias, from a design point of view, one has to follow the following steps:

1) Set f_{High}, the ending frequency of the bandgap.
2) Compute d by using (2.12).
3) Locate the stitching vias maintaining their distance less than d.

The approach introduced in this section, together with the conventional planar EBG design strategies already proposed, allows a quantitative design of the via location and patch geometry of an embedded EBG structure.

2.6.2 Introducing the Stitching Vias

The structure in Figure 2.17 is modified, and more stitching vias following a regular pattern are added. Two new configurations are considered having 25 and 41 stitching vias, respectively. The top views of these configurations are shown in Figure 2.21. The transfer functions $|S_{21}|$ between the two ports shown in Figure 2.21 are computed for the following configurations without the EBG layer: no stitching vias, 13, 21, and 41 stitching vias. The results are compared in Figure 2.22. The first resonant mode for the three cases with 13, 25, and 41 shorting vias occurs at 3.05, 4.56, 6.91 GHz, respectively. Table 2.2 associates the lowest resonant frequency f_{res} and the maximum distance d_{max} between two adjacent vias with the number of stitching vias. Furthermore, the distance d is computed by using (2.12) and the relative error between d_{max} and d is provided. The maximum error is always less than 20% and is less than 10% when the via pattern is a centered-like lattice (13 and 41 via cases).

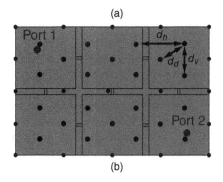

Figure 2.21 Top view of the embedded EBG layer with (a) 25 shorting vias and (b) 41 shorting vias. (Taken from Ref. [39]. Reproduced with permission of IEEE.)

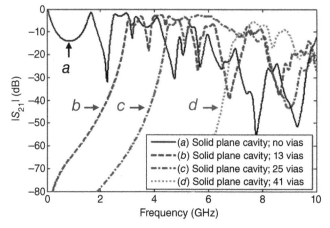

Figure 2.22 $|S_{21}|$ of the solid plane cavity. (a) Without vias. (b) With 13 shorting vias. (c) With 25 shorting vias. (d) With 41 shorting vias. (Taken from Ref. [39]. Reproduced with permission of IEEE.)

Table 2.2 Number of stitching vias, resonant frequency, and vias distance.

	13 Vias	25 Vias	41 Vias
f_{res} (GHz)	3.05	4.56	6.91
d_{max} (mm)	21.85 (d_h)	13.07 (d_d)	10.92 (d_h)
d (mm)	23.44	15.68	10.35
Error (%)	7.3	19.9	−5.2

The EBG layer is then included between the two solid planes; the resulting embedded EBG structure is excited as in Figure 2.16. Figures 2.23a and b compare $|S_{21}|$ for the cases with 25 and 41 stitching vias. The two $|S_{21}|$ of the embedded EBG structures in Figure 2.23 are equal below the bandgap region (black curves in Figure 2.23a and b), up 2.0 GHz, and have a similar trend in the first part of the bandgap. The bandgap ends at the first resonance of the solid plane cavity, at 4.56 and 6.91 GHz for the case of 25 vias (Figure 2.23a) and 41 stitching vias (Figure 2.23b), respectively.

Above the bandgap, independent of the number of stitching vias, $|S_{21}|$ of the embedded EBG structure strictly follows $|S_{21}|$ of the solid planes; thus, the two solid plane cavity, in effect, guide the behavior of the overall structure. The attenuated peak, at around 5 GHz in Figure 2.23b, is due to the single patch resonance (the TM_{01} mode of the single patch) [43].

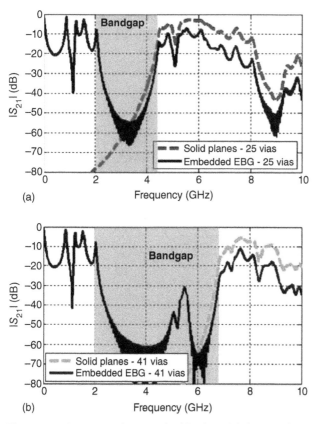

Figure 2.23 Comparison between $|S_{21}|$ for the solid planes and embedded EBG. (a) Case of 25 shorting vias. (b) Case of 41 shorting vias. (Taken from Ref. [39]. Reproduced with permission of IEEE.)

Its amplitude is lower than in the original planar case shown in Figure 2.13c. This phenomenon is explained by Figure 2.24a and b, where the spatial distribution of $|E_z|$ for a planar EBG (not embedded and without vias) and for an embedded EBG with 41 shorting vias are illustrated. The presence of the stitching vias, in the model in Figure 2.24b, alters the $|E_z|$ field distribution reducing the electromagnetic energy propagating from port 1 to port 2.

Some test boards are designed and manufactured in order to validate the design technique presented in this section. The planar dimensions of these boards are those shown in Figure 2.13a; however, they have a 0.2 mm dielectric thickness above and below the EBG layer. The thickness of the three metal planes is 0.017 mm. The stitching vias have a 0.3 mm diameter; the etched circular gap between vias and the patch at the EBG layer has a 1.27 mm diameter. The nominal relative dielectric constant and loss tangent of the FR-4

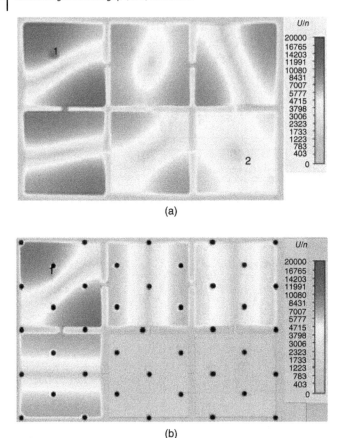

Figure 2.24 Spatial distribution of $|E_z|$ at 5.0 GHz. (a) Planar EBG (not embedded, no vias). (b) Embedded EBG with 41 shorting vias. (Taken from Ref. [39]. Reproduced with permission of IEEE.)

substrate employed in the PCB are $\varepsilon_r = 4.4$. and *tan* $\delta = 0.02$. A second-order Debye model has been used in order to represent the dispersive behavior of this dielectric material ($\varepsilon_\infty = 4.0$, $\varepsilon_{s1} = 4.1$, $\varepsilon_{s2} = 4.4$, $\tau_1 = 30$ ps, $\tau_2 = 10$ ps).

Two test samples with 25 regularly spaced vias, associated with the model in Figure 2.21a and included in the manufactured test board, are measured according to the setup in Figure 2.25. The first one is without the embedded EBG plane and the second with the patterned plane.

Two SMA connectors are mounted on each board at the port locations identified in Figure 2.13a. The *S*-parameters of the two test boards were measured using a 10 MHz–9 GHz VNA (Anritsu MS4624B). The measured $|S_{21}|$ are given in Figure 2.26a. The trend of the measured data is similar to the

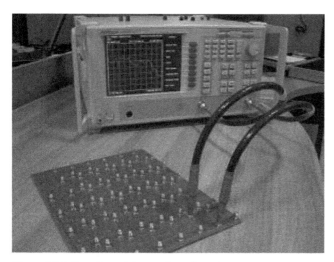

Figure 2.25 Overview of the test boards and measurement setup. (Taken from Ref. [39]. Reproduced with permission of IEEE.)

trend illustrated in Figure 2.23a. The embedded EBG has a resonant behavior that occurs below 2 GHz; this is due to the EBG solid plane cavity.

It is evident that the two curves in Figure 2.26a overlap in the range of 3–3.75 GHz up to the first resonance of the solid plane cavity (even though the resonant peak at 3.75 GHz is not clearly visible in the solid plane case associated with the red dashed curve). These measured data confirm the physical mechanism regulating the embedded EBG structure behavior previously introduced.

A simulation model associated with the manufactured embedded EBG structure with 25 vias was built in order to further validate these results. Figure 2.26b illustrates the comparison between the measured results and those obtained through three-dimensional simulation. The agreement between measured and simulated data is quantified in Figure 2.27 by using the feature selective validation (FSV) technique [44,45], according to the IEEE Standard 1597.1 [46]. Both the ADMc and parameters highlight the good agreement between the measured and simulated data, along with their associated values of grade and spread.

The use of planar EBG structures embedded in an inner layer of the PCB stack-up is both a viable and effective solution for noise mitigation. However, the design of an embedded EBG structure requires the presence of stitching vias that short the solid planes above and below the EBG layer. These vias inhibit the low-frequency resonances of the cavity made by the upper and lower solid planes. Since f_{High} is the high frequency limit of the desired bandgap, the vias can be located at a distance d such that (2.12) holds. This approach requires that the

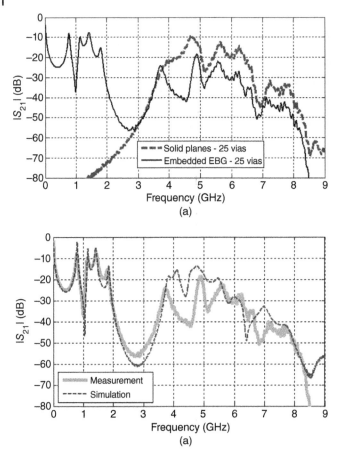

Figure 2.26 (a) Comparison between measured $|S_{21}|$ for the solid planes and embedded EBG for the case of 25 shorting vias. (b) Comparison between measured and simulated $|S_{21}|$ for an embedded EBG with 25 regularly placed shorting vias. (Taken from Ref. [39]. Reproduced with permission of IEEE.)

shorting vias are placed following a regular pattern in order to achieve an effective bandgap on the whole EBG area.

2.7 Application Examples

2.7.1 Splitting Power Planes by EBG Barrier

The previous analysis has been applied to a specific case of a two-layer power bus as part of a multilayer PCB. Typical uses of the EBG are mainly focused on

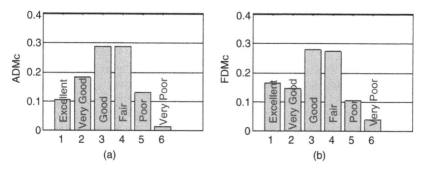

Figure 2.27 Feature selective validation techniques results for the comparison in Figure 2.26. (a) ADMc: grade = 3, spread = 4. (b) FDMc: grade = 3, spread = 4. (Taken from Ref. [39]. Reproduced with permission of IEEE.)

the layout of a complete plane. However, other layout constraints can limit the EBG design on the entire plane. The example presented herein introduces the possibility to design the EBG only on a limited plane section to be freely laid out where it is more appropriate without affecting other system functionalities. This application example of the EBG geometry is aimed at isolating two sections of the PCB, Area 1 and Area 2, where two high-speed ICs are located, IC_1 and IC_2, as shown in Figure 2.28. These ICs generate high-speed digital signals at a data rate of 10 Gb/s, thus considering the possible noise source related to a band centered at the data rate fundamental harmonic, 5 GHz. The important geometry features of the overall board are shown in Figure 2.28. The dielectric between the two metal planes is 0.4 mm thick; it has a dielectric permittivity $\varepsilon_r = 4.4$.

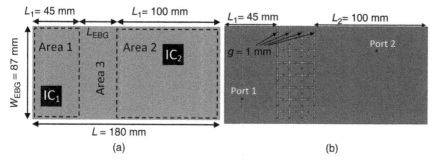

Figure 2.28 (a) Geometry of the two-layer PCB considered in the example. (b) Design of the 3 × 8 EBG to fit the unconstrained Area 3. (Taken from Ref. [34]. Reproduced with permission of The Electromagnetics Academy.)

The lower GND plane is kept solid, whereas the top plane, PWR, is employed to include an EBG geometry. Limiting the EBG area can avoid problems related to signal integrity (i.e., a solid plane is always preferred as signal reference). Therefore, the limited EBG portion can be laid out wherever it is more appropriate according to other constraints, that is, in Area 3 in Figure 2.28, with the following size: $L_{EBG} = 35$ mm and $W_{EBG} = 87$ mm.

The size of the EBG analyzed in Section 2.3 ($a = b = 10$ mm, $w = 0.5$ mm, $g = 1$ mm) can achieve a bandgap centered at approximately 5 GHz; these dimensions will be considered for designing the EBG for the Area 3. According to the size of Area 3, a 3×8 EBG matrix of patches could fully cover this plane portion, with overall size 35 mm × 87 mm. In the case of Area 1 and Area 2 having associated the same voltage level, the DC connection between the two areas could be preferable; thus, the EBG section can be designed to be connected to the two areas through bridges with the same size as the EBG bridges. Besides the large 3×8 configuration, a second EBG geometry is designed with 1×8 array of patches, keeping just the second row of the 3×8 configuration. This second model leads to make the Area 1 and Area 2 larger, with $L_1 = 57$ mm and $L_2 = 111$ mm. The two EBG geometries are first simulated separately, as standalone EBG cavities. The simulation results are shown in Figure 2.29 together with $|S_{21}|$ of the solid plane geometry. The prediction of the bandgap f_{Low} by the proposed analytical procedure provides $f_{Low} = 3.7$ GHz and $f_{Low} = 4.85$ GHz for the case 1×8 and 3×8 EBG, respectively. These results agree well with the last visible resonant mode before the bandgap (3.34 and 4.46 GHz, respectively) identified in the solid and dashed curves in Figure 2.29. Also the f_{High} from the 3D simulation equal to 7.06 GHz

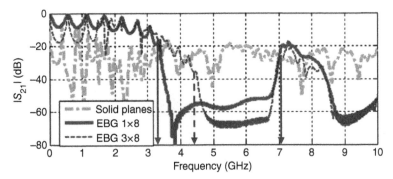

Figure 2.29 $|S_{21}|$ for the geometry made by two solid planes, the 1×8 and the 3×8 EBG structures. (Taken from Ref. [34]. Reproduced with permission of The Electromagnetics Academy.)

is correctly predicted by (2.2) that gives 7.15 GHz. The simulation ports in the large plane model are placed at the locations of IC_1 and IC_2. Although the two EBG structures have different bandgap lower limits, they both include a band around 5 GHz. The 1×8 structure provides a more robust design with a bandgap going from 3.34 to 7.06 GHz. The 3×8 geometry, instead, with a $f_{Low} = 4.46$ GHz, has a bandgap not centered around 5 GHz; however, it is more effective providing larger noise reduction. The two EBG configurations are included in the full model, as in Figure 2.28b for the 3×8 case.

The EBG geometry designed as proposed in Figure 2.28 achieves a sort of fence isolating the two areas, below the bandgap, keeping a similar behavior as the solid plane case. This can be seen in the simulation results in Figure 2.30 and in the $|Z_{11}|$ curve plotted in Figure 2.31. The small shift, in the $|Z_{11}|$ toward lower frequencies of the first resonant mode, from 400 to 360 MHz and to 320 MHz, for the 1×8 and the 3×8 case, respectively, can be managed by accordingly selecting and placing decoupling capacitors [47]. The better isolation between the two areas is obtained with the implementation of the 3×8 EBG geometry that achieves the largest reduction in $|S_{21}|$ from around 4 to 7 GHz. This 3×8 EBG provides about 10–20 dB of noise reduction with respect to the original solid plane case, and 5–10 dB with respect to the 1×8 EBG case. Thus, the EBG with three rows (3×8) is more effective than the 1×8 EBG case, if no constraints on layout space force to use the 1×8 EBG.

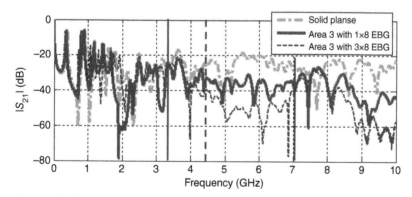

Figure 2.30 $|S_{21}|$ for the geometry made by two solid planes, and the planes after including the 1×8 or the 3×8 EBG structure between Areas 1 and 2. (Taken from Ref. [34]. Reproduced with permission of The Electromagnetics Academy.)

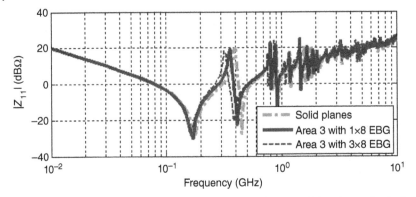

Figure 2.31 $|Z_{11}|$ for the geometry made by two solid planes, and the planes after including the 1 × 8 and the 3 × 8 EBG structures. (Taken from Ref. [34]. Reproduced with permission of The Electromagnetics Academy.)

2.7.2 IC Isolation by an EBG Fence

A multilayer PCB that includes several high-speed ICs is presented as a second application example. One of the ICs operates at 10 Gbps, with the first harmonic at 5 GHz. The noise generated by this IC can propagate within the power planes; thus, it needs to be reduced over a frequency range as wide as possible that includes, at minimum, the 5 GHz fundamental harmonic. The PCB area of interest is centered at the IC location, as from Figure 2.32a, where the PCB excitation port is placed. The dashed line identifies the IC outline having a side length of 30 mm. The overall PCB area considered in the study is 80 mm × 80 mm. The Area 1 contains the IC and a ring to allow the placement of local decoupling capacitors. A more external ring (Area 2) is reserved for the EBG layout to realize a sort of fence around the IC. The remaining PCB area is limited to the additional external ring (Area 3) needed for the placement of the receiving port (Port 2) chosen to be 10 mm wide. A larger overall area does not add any information on the study of the EBG effectiveness. An optional port location is considered (Port 1') to evaluate the effect of the distance between the source and receiving port when divided by the same EBG fence. The stack-up needed for the EBG design consists of two metal layers embedding a 0.2 mm dielectric characterized by $\varepsilon_r = 4.4$.

Area 2 is divided into four parts (Parts 1–4) to place four equal planar EBG structures. The EBG is designed according to the quick and efficient procedure detailed in Refs [13,17]. The size *a* of the EBG patch fixes the bandgap upper limit f_{High} according to (2.2), as the first resonant mode TM_{10} of the single patch.

The patch size is fixed to be $a = 5$ mm, providing a theoretical value of f_{High} = 14.3 GHz. The design of the other EBG parameters such as the bridge width w, bridge length g, and number of patches M will affect the bandgap lower limit f_{Low}. The bandgap lower limit should be lower than 5 GHz, according to the noise source spectrum, as mentioned above. The f_{Low} of an $M \times 1$ EBG array is associated with the resonant mode $TM_{M-1,0}$ of the overall patterned EBG structure. The value of 3.72 GHz, lower than 5 GHz, can be achieved by selecting $M = 4$, $w = 0.2$ mm, $g = 1.5$ mm. The limited length of the EBG ($4a + 3g = 24.5$ mm) allows us to place two EBGs in each Part of Area 2. A possible design is reported in Figure 2.32b where the ring for the decoupling capacitor placement is 7 mm wide, and the Area 2 width is calculated to be 8 mm ($2g + a$). Two 4×1 EBGs are placed instead of one 8×1 array for keeping the f_{Low} as small as possible (with constant a, w, and g, the f_{Low} increases by increasing M), even though both layout options could be used to fit the Area 2 surface. Therefore, the two adjacent EBGs are not connected through the bridge, as detailed in Figure 2.32b. Each patch of the EBG is connected to Areas 1 and 3 using the same bridges as for the EBG itself.

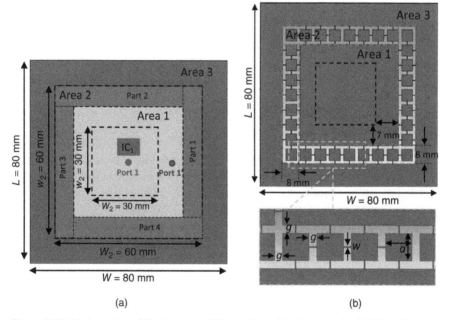

(a)   (b)

Figure 2.32 (a) Geometry of the two-layer PCB considered in the example. (b) PCB with the implemented EBG and details of the 4×1 EBG. (Taken from Ref. [48]. Reproduced with permission of IEEE.)

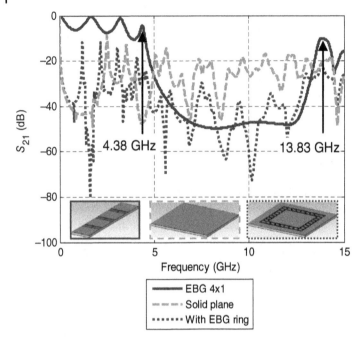

Figure 2.33 Comparison between $|S_{21}|$ of the 4 × 1 EBG, solid plane case and with the addition of the EBG ring. (Taken from Ref. [48]. Reproduced with permission of IEEE.)

A preliminary simulation is run for the standalone 4 × 1 EBG. The results are shown in Figure 2.33 (solid curve). The simulated f_{Low} is 4.38 GHz, larger than the designed values of 3.72 GHz (the error in the analytical procedure is larger for smaller number of patches). The $f_{High} = 13.83$ GHz is smaller than the theoretical value obtained from (2.2) the error is below 10% as expected from $g/a = 0.3$, as described in Figure 2.7. Figure 2.33 also shows the $|S_{21}|$ results related to a 80 × 80 mm solid plane (dashed curve) and the structure with the EBG in Figure 2.32b (dotted curve). A clear $|S_{21}|$ reduction is obtained in the frequency range corresponding to the bandgap, from 4.75 to 13.6 GHz. Moving from the solid plane case to the case with the EBG fence, the insertion loss attenuation is about 10–20 dB in the whole frequency range of interest.

An additional consideration involves the ability of the power planes, where the EBG is laid out, to deliver the correct amount of current from the VRM to the IC. It is known that the parasitic inductance of the voltage regulator module (VRM), decaps, planes, IC pin vias, and so on limits the supply currents at different frequency bands, depending on their inductance values (0.01–1 µH for VRM and bulk capacitors, 0.1–10 nH for SMT decaps, and

1–100 pH for on-package and on-chip decaps) [49]. The bridge inductance associated with a single EBG bridge is, in this particular case, 630 pH; this value should be doubled to account for the two bridges connecting Areas 1 and 3 for each EBG patch. However, taking into account the seven bridge pairs connecting Areas 1 and 3 along each side, a rough but reasonable calculation lead to $L_{EBG\text{-}bridge} = 180$ pH additional parasitic inductance given by the narrow EBG bridges (taking into account all four sides will lead to an even lower value). The associated impedance $|Z_{EBG\text{-}bridge}| = \omega \cdot L_{EBG\text{-}bridge}$ goes beyond 10 mΩ at about 10 MHz. Therefore, the EBG bridge inductance is quite lower than the parasitic inductance of VRM and bulk capacitances usually placed far from the IC, $L_{EBG\text{-}bridge}$, and is comparable to the parasitic values of low-inductance SMT decap. A good compromise is to place low-inductance decaps close to the IC power pins, inside Area 1, and standard SMT decaps in Area 3 or directly on the EBG patch.

Additional considerations related to the applicability of this EBG fence surrounding the IC within the multilayer stack-up are not discussed here; however, a detailed layout technique is shown in [48].

2.8 Conclusions

This chapter introduces a complete procedure for the analysis of planar electromagnetic bandgap structures based on the concept of total inductance. A patterned EBG plane embedded together with a solid plane underneath generates a cavity that has more inductance than the ideal cavity counterpart made by two solid planes. This procedure allows a quick and accurate evaluation of the bandgap lower limit f_{Low}. The study also provides a more precise evaluation of the bandgap upper limit f_{High} quantifying the effect of the parameters altering its analytical calculation. The value of the f_{Low} is studied as a function of the number of patches, finding out that it reaches an asymptotic value when the number of patches goes toward infinity, while keeping constant the geometrical and electrical properties of the structure. This limit allows a quick evaluation of f_{Low}. Then the analytical expression of f_{Low} gives a more precise calculation taking into account the number of patches chosen according to the available layout area. The $M \times 1$ array achieves a lower f_{Low} rather than the $M \times N$ case and thus a wider bandgap due to the constant f_{High}. The $M \times N$ configuration, instead, could be more helpful when using the EBG for a partial filling of the plane, achieving better noise isolation along both directions, as demonstrated in the practical example. The layout option of partial EBG has been shown to be effective for reducing noise coupling between two PCB areas [50–59], without the need of patterning the whole plane, thus keeping the ideal solid reference where required by other constraints.

References

1 S.V. Berghe, F. Olyslager, D. de Zutter, J.D. Moerloose, and W. Temmerman, Study of the ground bounce caused by power plane resonances. *IEEE Trans. Electromagn. Compat.*, Vol. 40, No. 2, 1998, pp. 111–119.

2 N. Na, J. Jinseong, S. Chun, M. Swaminathan, and J. Srinivasan, Modeling and transient simulation of planes in electronic packages. *IEEE Trans. Adv. Packaging*, Vol. 23, No. 3, 2000, pp. 340–352.

3 Z.L. Wang, O. Wada, Y. Toyota, and R. Koga, Analysis of resonance characteristics of a power bus with rectangle and triangle elements in multilayer PCBs, in *Proc. of the Asia-Pacific Conference on Environmental Electromagnetics*, Nov. 2003, pp. 73–76.

4 G.T. Lei, R.W. Techentin, P.R. Hayes, D.J. Schwab, and B.K. Gilbert, Wave model solution to the ground/power plane noise problem. *IEEE Trans. Instrum. Meas.*, Vol. 44, No. 2, 1995, pp. 300–303.

5 T. Okoshi, Chapter 3 in *Planar Circuits for Microwaves and Lightwaves*, Springer, Berlin, Germany, 1985.

6 Z.Z. Oo, E.X. Liu, E.P. Li, X. Wei, Y. Zhang, M. Tan, L.W.J. Li, and R. Vahldieck, A semi-analytical approach for system-level electrical modeling of electronic packages with large number of vias. *IEEE Trans. Adv. Packaging*, Vol. 31, No. 2, 2008, pp. 267–274.

7 J. Kim, Y. Jeong, J. Kim, J. Lee, C. Ryu, J. Shim, M. Shin, and J. Kim, Modeling and measurement of interlevel electromagnetic coupling and fringing effect in a hierarchical power distribution network using segmentation method with resonant cavity model. *IEEE Trans. Adv. Packaging*, Vol. 31, No. 3, 2008, pp. 544–557.

8 W.-T. Huang, C.-H. Lu, and D.-B. Lin, The optimal number and location of grounded vias to reduce crosstalk. *Prog. Electromagn. Res.*, Vol. 95, 2009, pp. 241–266.

9 B. Wu and L. Tsang, Full-wave modeling of multiple vias using differential signaling and shared antipad in multilayered high speed vertical interconnects. *Prog. Electromagn. Res.*, Vol. 97 2009, pp. 129–139.

10 A.E. Engin, K. Bharath, M. Swaminathan, M. Cases, B. Mutnury, N. Pham, D.N. de Araujo, and E. Matoglu, Finite-difference modeling of noise coupling between power/ground planes in multilayered packages and boards, in *Proc. of the 56th Electronic Components and Technology Conference*, June 2006, pp. 1262–1267.

11 J. Choi, V. Govind, R. Mandrekar, S. Janagama, and M. Swaminathan, Noise reduction and design methodology for the mixed-signal systems with

alternating impedance electromagnetic bandgap (Al-EBG) structure, in *2005 IEEE MTT-S International Microwave Symposium Digest, Long Beach, CA,* June 2005, pp. 645–651.

12 T.H. Kim, D. Chung, E. Engin, W. Yun, Y. Toyota, and M. Swaminathan, A novel synthesis method for designing electromagnetic bandgap (EBG) structures in packaged mixed signal systems, in *Proc. of the 56th Electronic Components and Technology Conference*, 2006, pp. 1645–1651.

13 J. Choi, V. Govind, and M. Swaminathan, A novel electromagnetic bandgap (EBG) structure for mixed-signal system applications, in *Proc. of the IEEE Radio Wireless Conference*, September 2004, pp. 243–246.

14 D. Sievenpiper, L. Zhang, R.F.J. Broas, N.G. Alexopolous, and E. Yablonovitch, High impedance electromagnetic surfaces with a forbidden frequency band. *IEEE Trans. Microw. Theory Tech.*, Vol. 47, No. 11, 1999, pp. 2059–2074.

15 S.M. Moghadasi, Amir Reza Attari, and Mir M. Mirsalehi, Design of three-layer circular mushroom-like EBG structures. *PIERS Online*, Vol. 4, No. 2, 2008, pp. 217–220.

16 S.M. Moghadasi, A.R. Attari, and M.M. Mirsalehi, Compact and wideband 1-D mushroom-like EBG filters. *Prog. Electromagn. Res.*, Vol. 83, 2008, pp. 323–333.

17 S.-H. Kim, T.T. Nguyen, and J.-H. Jang, Reflection characteristics of 1-D EBG ground plane and its application to a planar dipole antenna. *Prog. Electromagn. Res.*, Vol. 120, 2011, pp. 51–66.

18 M. Hajj, R. Chantalat, M. Lalande, and B. Jecko, Sectoral M-EBG antenna with multipolarization capabilities for Wimax base stations. *Prog. Electromagn. Res. C*, Vol. 22, 2011, pp. 211–229.

19 I. Khromova, I. Ederra, R. Gonzalo, and B.P. de Hon, Symmetrical pyramidal horn antennas based on EBG structures. *Prog. Electromagn. Res. B*, Vol. 29, 2011, pp. 1–22.

20 H.-H. Xie, Y.-C. Jiao, L.-N. Chen, and F.-S. Zhang, An effective analysis method for EBG reducing patch antenna coupling. *Prog. Electromagn. Res. Lett.*, Vol. 21, 2011, pp. 187–193.

21 T.H. Hubing, J.L. Drewniak, T.P. Van Doren, and D.M. Hockanson, Power-bus decoupling on multi-layer printed circuit boards. *IEEE Trans. Electromagn. Compat.*, Vol. 37, 1995, pp. 155–166.

22 R. Abhari and G.V. Eleftheriades, Metallo-dielectric electromagnetic bandgap structures for suppression and isolation of the parallel-plate noise in high-speed circuits. *IEEE Trans. Microw. Theory Tech.*, Vol. 51, No. 6, 2003, pp. 1629–1639.

23 D.-S. Eom., J. Byun, and H.-Y. Lee, New composite power plane using spiral EBG and external magnetic material for SSN suppression. *Prog. Electromagn. Res. Lett.*, Vol. 15, 2010, pp. 69–77.

24 H.-S. He, X.-Q. Lai, Q. Ye, Q. Wang, W.-D. Xu, J.-G. Jiang, and M.-X. Zang, Wideband SSN suppression in high-speed PCB using novel planar EBG. *Prog. Electromagn. Res. Lett.*, Vol. 18, 2010, pp. 29–39.

25 A. Tavallaee and R. Abhari, 2-D characterization of electromagnetic bandgap structures employed in power distribution networks. *IET Microw. Antennas Propag.*, Vol. 1, 2007, pp. 204–211.

26 A. Tavallaee, M. Iacobacci, and R. Abhari, A new approach to the design of power distribution networks containing electromagnetic bandgap structures, in *2006 IEEE Electrical Performance of Electrical Packaging*, 2006, pp. 43–46.

27 T. Kamgaing and O. Ramahi, High-impedance electromagnetic surfaces for parallel-plate mode suppression in high-speed digital systems, in *2002 IEEE Topical Meeting on Electrical Performance of Electrical Packaging*, 2002, pp. 279–282.

28 T. Kamgaing and O.M. Ramahi, Design and modeling of high impedance electromagnetic surfaces for switching noise suppression in power planes. *IEEE Trans. Electromagn. Compat.*, Vol. 47, No. 3, 2005, pp. 479–489.

29 T.-K. Wang, T.W. Han, and T.-L. Wu, A novel power/ground layer using artificial substrate EBG for simultaneously switching noise suppression. *IEEE Trans. Microw. Theory Tech.*, Vol. 56, No. 5, pp. 2008, pp. 1164–1171.

30 F. de Paulis, L. Raimondo, S. Connor, B. Archambeault, and A. Orlandi, Compact configuration of a planar EBG based CM filter and crosstalk analysis, in *Proc. of the IEEE International Symposium on EMC*, August 14–19, 2011, Long Beach, CA.

31 B. Mohajer-Iravani and O.M. Ramahi, Wideband circuit model for planar EBG structures. *IEEE Trans. Adv. Packaging*, Vol. 33, No. 2, 2010, pp. 345–354.

32 Y. Toyota, E. Engin, T. Kim, M. Swaminathan, and K. Uriu, Stopband prediction with dispersion diagram for electromagnetic bandgap structures in printed circuit boards, in *Proc. of the IEEE EMC Symposium, Portland, OR*, August 2006, pp. 807–811.

33 K.H. Kim and J.E. Schutt-Aine, Analysis and modeling of hybrid planar type electromagnetic-bandgap structures and feasibility study on power distribution network applications. *IEEE Trans. Microw. Theory Tech.*, Vol. 56, No. 1, 2008, pp. 178–186.

34 F. de Paulis and A. Orlandi, Accurate and efficient analysis of planar electromagnetic band-gap structures for power bus noise

mitigation in the GHz band. *Prog. Electromagn. Res. B*, Vol. 37, 2012, pp. 59–80.

35 F. de Paulis, L. Raimondo, and A. Orlandi, IR-Drop analysis and thermal assessment of planar electromagnetic band-gap structures for power integrity applications. *IEEE Trans. Adv. Packaging*, Vol. 33, No. 3, 2010, pp. 617–622.

36 L. Raimondo, F. de Paulis, and A. Orlandi, A simple and efficient design procedure for planar electromagnetic bandgap structures on printed circuit boards. *IEEE Trans. Electromagn. Compat.*, Vol. 53, No. 2, 2010, pp. 482–490.

37 Computer Simulation Technology , CST Studio Suite, 2011 (online). Available at http://www.cst.com/

38 T.H. Kim, D. Chung, E. Engin, W. Yun, Y. Toyota, and M. Swaminathan, A novel synthesis method for designing electromagnetic band gap (EBG) structures in packaged mixed signal systems, in *Proc. of the 56th Electronic Components and Technology Conference, San Diego, CA*, May 30–June 2, 2006, pp. 1645–1651.

39 F. de Paulis, L. Raimondo, and A. Orlandi, Impact of shorting vias placement on embedded planar electromagnetic bandgap structures within multilayer printed circuit boards. *IEEE Trans. Microw. Theory Tech.*, Vol. 58, No. 7, 2010, pp. 1867–1876.

40 S. Huh, M. Swaminathan, and F. Muradali, Design, modeling, and characterization of embedded electromagnetic band gap (EBG) structure, in *Proc. of the IEEE Electrical Performance of Electronic Packaging Conference*, October 2008, pp. 83–86.

41 T.L. Wu, Y.H. Lin, T.K. Wang, C.C. Wang, and S.T. Chen, Electromagnetic bandgap power/ground planes for wideband suppression of ground bounce noise and radiated emission in high-speed circuits. *IEEE Trans. Microw. Theory Tech.*, Vol. 53, No. 9, 2005, pp. 2935–2942.

42 K. Payandehjoo, A. Tavallaee, and R. Abhari, Analysis of shielded electromagnetic bandgap structures using multiconductor transmission-line theory. *IEEE Trans. Adv. Packaging*, Vol. 33, No. 1, 2010, pp. 236–245.

43 T.-L. Wu, H.-H. Chuang, and T.-K. Wang Overview of power integrity solutions on package and PCB: decoupling and EBG isolation. *IEEE Trans. Electromagn. Compat.*, Vol. 52, No. 2, 2010, pp. 346–356.

44 A.P. Duffy, A.J.M. Martin, A Orlandi, G Antonini, T.M. Benson, and M.S. Woolfson, Feature selective validation (FSV) for validation of computational electromagnetics (CEM). Part I: the FSV method. *IEEE Trans. Electromagn. Compat.*, Vol. 48, No. 3, 2006, pp. 449–459.

45 A Orlandi, A.P. Duffy, B. Archambeault, G. Antonini, D.E. Coleby, and S. Connor, Feature Selective Validation (FSV) for validation of computational electromagnetics (CEM). Part II: assessment of FSV performance. *IEEE Trans. Electromagn. Compat.*, Vol. 48, No. 3, 2006, pp. 460–467.

46 IEEE Standard P1597,Standard for Validation of Computational Electromagnetics Computer Modeling and Simulation: Part 1 and Part 2, IEEE, 2008.

47 F. de Paulis, F. Vasarelli, and A. Orlandi et al., IR-drop calculation in EMStudio, CST Application Notes, 2009.

48 F. de Paulis, M.N. Nisanci, and A. Orlandi, Practical EBG application to multilayer PCB: impact on power integrity. *IEEE Electromagn. Compat. Mag.*, Vol. 1, No. 3, 2012, pp. 60–65.

49 L.D. Smith, M Sarmiento, Y Tretiakov, S. Sun, Z. Li, and S. Chandra, PDN resonance calculator for chip, package and board, in *Proc. of the IEC DesignCon 2012*, January 30–February 2, 2012, Santa Clara, CA.

50 A. Mahmoudian and J.A. Rashed-Mohassel, Reduction of EMI and mutual coupling in array antennas by using DGS and AMC structures. *PIERS Online*, Vol. 4, No. 1, 2008, pp. 36–40.

51 T.L. Wu, Y.H. Lin, T.K. Wang, C.C. Wang, and S.T. Chen, Electromagnetic bandgap power/ground planes for wideband suppression of ground bounce noise and radiated emission in high-speed circuits. *IEEE Trans. Microw. Theory Tech.*, Vol. 53, No. 9, 2005, pp. 2935–2942.

52 S. Shahparnia and O.M. Ramahi, Electromagnetic interference (EMI) reduction from printed circuit boards (PCB) using electromagnetic bandgap structures. *IEEE Trans. Electromagn. Compat.*, Vol. 46, No. 4, 2004, pp. 580–587.

53 B. Mohajer-Iravani, S. Shahparnia, and O.M. Ramahi, Coupling reduction in enclosures and cavities using electromagnetic band gap structures. *IEEE Trans. Electromagn. Compat.*, Vol. 48, No. 2, 2006, pp. 292–303.

54 F. de Paulis, A. Orlandi, L. Raimondo, B. Archambeault, and S. Connor, Common mode filtering performances of planar EBG structures, in *Proc. of the IEEE International Symposium on Electromagnetic Compatibility*, 2009, pp. 86–90.

55 F. de Paulis, L. Raimondo, S. Connor, B. Archambeault, and A. Orlandi, Design of a common mode filter by using planar electromagnetic band-gap structures. *IEEE Trans. Adv. Packaging*, Vol. 33, No. 4, 2010, pp. 994–1002.

56 D.M. Pozar, *Microwave Engineering*, 3rd ed. John Wiley & Sons, Inc., New York, 2005.

57 J. Fan, J.L. Drewniak, J.L. Knighten, N.W. Smith, A. Orlandi, T.P. Van Doren, T.H. Hubing, and R.E. DuBroff, Quantifying SMT decoupling capacitor placement in DC power-bus design for multilayer PCBs. *IEEE Trans. Electromagn. Compat.*, Vol. 43, No. 4, 2001, pp. 588–599.

58 Institute for Interconnecting and Packaging Electronic Circuits, IPC Standard IPC-A-600-3.2.4 "Acceptability of Printed Boards", July 2004.

59 S. Uka, M. Ilami, and N. Chia, Measurement techniques for DC resistance. *IEEE Trans. Instrum. Meas.*, Vol. 53, No. 5, pp. 1392–1396.

3

Impact of Planar Ebgs on Signal Integrity in High-Speed Digital Boards

The aim of this chapter is to investigate the electromagnetic phenomena involved in the propagation of a signal on a microstrip over an EBG structure when an interconnect is referenced to the shaped (or patterned) layer. This analysis required correct and reliable use of the EBG structures in a real-world PCB design. We will study the coupling between the electromagnetic field associated with the transmitted signal and the field inside the cavity over a large frequency range. In Section 3.2, the behavior of the TM_z (z being the vertical axis perpendicular to the planes) cavity resonances for an EBG structure built up by rectangular patches connected by straight bridges is investigated and compared with the behavior of an ideal cavity with a solid plane. This Power Integrity (PI) analysis is carried out up to 6 GHz by using a finite integration technique-based numerical tool [1]. Then the coupling, and the associated effects, between the electromagnetic field due to the microstrip and the EBG cavity are studied. This coupling occurs through the gaps and slots of the patterned plane. This Signal Integrity (SI) analysis results in the resonances in the cavity that are related to those of the S-parameters of the signal interconnect. The relationship between the power plane impedance profile and the transfer function of the microstrip line is investigated. A correspondence between the SI and the PI response is found and an explanation of the electromagnetic phenomena involved is given. Several different configurations are studied to validate the developed concepts. This chapter is based on the work in Ref. [2] for Section 3.1, and in Ref. [3] for Section 3.2.

3.1 Coupling Mechanisms Between Microstrip Lines and Planar Ebgs

In a real board where the EBG structure is employed in the stack-up of a multilayer PCB, some traces will be referenced to the EBG patterned plane. In

Electromagnetic Bandgap (EBG) Structures: Common Mode Filters for High-Speed Digital Systems,
First Edition. Antonio Orlandi, Bruce Archambeault, Francesco De Paulis, and Samuel Connor.
© 2017 by The Institute of Electrical and Electronics Eingineers, Inc. Published 2017 by John Wiley & Sons, Inc.

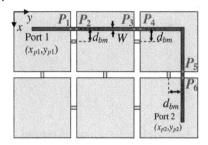

Figure 3.1 Layout of the microstrip on top of the EBG plane. (Taken from Ref. [2]. Reproduced with permission of IEEE.)

this section, the behavior of the transfer function $|S_{21}^{SI}|$ of an L-shaped microstrip routed as in Figure 3.1 over the EBG is considered. The microstrip runs from Port 1 to Port 2, it is $w = 0.4$ mm wide, has a nominal impedance of 50 Ω, and it is at $h = 0.2$ mm above the shaped plane. The dielectric between the EBG plane and the microstrip is the same as the dielectric inside the EBG cavity (i.e., between the EBG patterned layer and the solid ground layer beneath it). In the first case, the distance between the bridge and the trace is $d_{bm} = 1.85$ mm (equal to $4.6 \cdot w$). In Figure 3.1, the locations $P_1, \ldots, P_6$ identify the points in which the signal return current meets the gaps.

$|S_{21}^{SI}|$ and $|S_{21}^{PI}|$ are compared in order to find the mutual coupling between the signal propagation on the microstrip and the EBG cavity behavior. This comparison is shown in Figure 3.2 where $|S_{21}|$ (dot–dashed curve) for the same interconnect routed over a solid plane.

As expected, the dot–dashed curve has the flat trend similar to an ideal lossy transmission line over a continuous plane. In the case of the strip routed over the patterned plane, notches occur in the microstrip $|S_{21}^{SI}|$ (dashed curve) whenever peaks occur in the EBG $|S_{21}^{PI}|$ (solid curve). The reverse is also

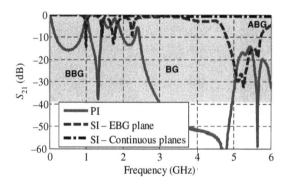

Figure 3.2 $|S_{21}|$ of the microstrip routed over a solid plane (dot–dashed curve), the EBG plane (dashed curve), and $|S_{21}^{PI}|$ (solid curve). The three regions are indicated: below bandgap (BBG), bandgap (BG), and above bandgap (ABG). (Taken from Ref. [2]. Reproduced with permission of IEEE.)

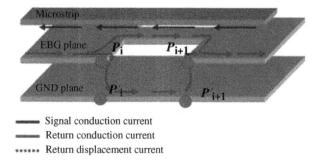

— Signal conduction current
— Return conduction current
•••••• Return displacement current

valid: flat and high (close to 0 dB) values of $|S_{21}^{SI}|$ corresponds to notches in $|S_{21}^{PI}|$.

The gaps in the patterned reference plane that are along the signal current return path allow energy coupling between the microstrip/EBG pair and the cavity made of the EBG and the continuous plane beneath it. The return current goes back to the source mainly flowing under the microstrip. The current, always choosing the path at least impedance, in the case of a discontinuity such as the gap along its path, can follow three different paths. The first is through the bridge (at LF due to the highly inductive nature of this path), another is through the patch-to-patch capacitance, and the other alternative is through the solid plane underneath the EBG patterned plane, as is shown in Figure 3.3.

The displacement current flowing from the EBG layer and the solid layer excites the cavity resonances at the frequencies in the below bandgap (BBG) region for the PI, as shown in Figure 3.2. Therefore, there is energy coupled between the microstrip and the cavity itself. No resonance can be excited within the EBG cavity at the SI bandgap (BG) region due to the deep (greater than −30 dB) and wide bandgap in the PI response. In order to better understand this mechanism, the impedances associated with the mentioned return paths are computed: $Z_{pp}(j\omega)$ defined by the pair of points P_i and P'_i ($i = 1, 2, \ldots, 6$; see Figures 3.1 and 3.3) looking into the cavity, and $Z_{ii}(j\omega)$ defined by the pair of points P_i and P_{i+1} ($i = 1, 3, 5$; see Figures 3.1 and 3.3) looking between the patches. Figure 3.4 reports the magnitude of both impedances.

Looking at the values of the impedance in the BG region ($f_{Low} = 2.6\,\text{GHz} < f < 5.1 = f_{High}$), $|Z_{pp}|$ is much lower than $|Z_{ii}|$ indicating that the preferred path of the microstrip return current is not through the gap between patches nor through the bridges, but on the continuous plane underneath the EBG patterned plane.

The microstrip behaves as if it were referenced to a continuous plane due to this low-inductance path; therefore, $|S_{21}^{SI}|$ is high (close to 0 dB) and flat (dashed

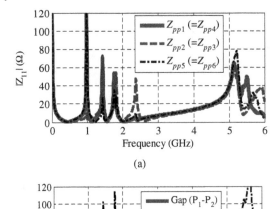

(a)

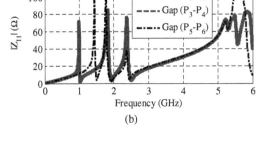

(b)

Figure 3.4 Frequency spectrum of (a) $|Z_{pp}|$ and (b) $|Z_{ii}|$ at different locations as in Figure 3.2. (Taken from Ref. [2]. Reproduced with permission of IEEE.)

curve in Figure 3.2). As $|Z_{pp}|$ increases (above 4 GHz), these positive properties of the return path are spoiled and $|S_{21}^{SI}|$ starts to decrease.

The same geometry as in Figure 3.1 has been simulated by removing the lower solid plane, reducing the geometry to a simple microstrip trace on a patterned plane (removing the EBG cavity structure). In Figure 3.5, $|S_{21}|$ of the microstrip is compared with and without the EBG cavity. In the latter case, the distributed behavior of the cavity is not present; therefore, no more resonances and bandgap are identified. The lack of the EBG cavity cannot help the signal propagation at the BG frequencies; therefore, the bridge inductance along the

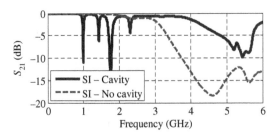

Figure 3.5 Frequency spectra of $|S_{21}|$ with EBG cavity (solid curve) and without cavity (dashed curve). (Taken from Ref. [2]. Reproduced with permission of IEEE.)

Figure 3.6 Frequency spectrum of $|S_{21}^{SI}|$: $d_{bm} = 0$ mm (solid curve), $d_{bm} = 1.85$ mm (dashed curve), and $d_{bm} = 4.85$ mm (dot–dashed curve). (Taken from Ref. [2]. Reproduced with permission of IEEE.)

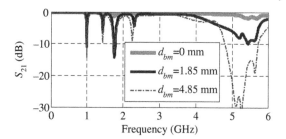

current return path limits the return current, thus worsening $|S_{21}|$ from 3 GHz on, as shown through the dashed curve in Figure 3.5.

Figure 3.6 shows $|S_{21}^{SI}|$ for three different distances d_{bm} of the microstrip from the bridges of the EBG patterned plane.

In the first case, the microstrip is routed exactly on top of the bridge ($d_{bm} = 0$ mm); the other two cases consider $d_{bm} = 1.85$ mm (equal to $5w$) and $d_{bm} = 4.85$ mm (equal to $13w$).

In BBG region, the bridge is further away from the microstrip, resulting in a longer and more inductive return path for the current, and hence the notches in the SI response are deeper. In the BG (around 2.6–5 GHz), since the return current does not significantly pass through the bridge, there is not much difference between the curves. The microstrip routed on top of the bridges has very small notches; the coupling between the top (the microstrip) and the bottom portion of the geometry (the EBG cavity) increases having a bigger impact on the microstrip transfer function when moving the trace away from the bridge; the notches are deeper for the case of $d_{bm} = 1.85$ mm and $d_{bm} = 4.85$ mm.

3.2 Impact of EBG Reference to Striplines

The basic planar EBG geometry is realized by a suitably designed patterned plane (see Figure 3.7a) on top of a continuous one, creating a cavity as illustrated in Figure 3.7b [2].

This configuration shows a bandgap on the frequency spectrum of $|S_{21}|$ between Port 1 (P_1) (5 mm, 5 mm) and Port 2 (P_2) (38.7 mm, 23.7 mm) from $f_{Low} = 2.7$ GHz to $f_{High} = 5$ GHz, as shown by the solid curve in Figure 3.8. If a third continuous layer is added on top of the EBG layer (as in Figure 3.7c), a new subcavity is obtained. For this new configuration, the spectrum of $|S_{21}|$ between the same two ports mentioned above shows distributed resonances along the whole frequency range of interest, as illustrated by the dashed curve in

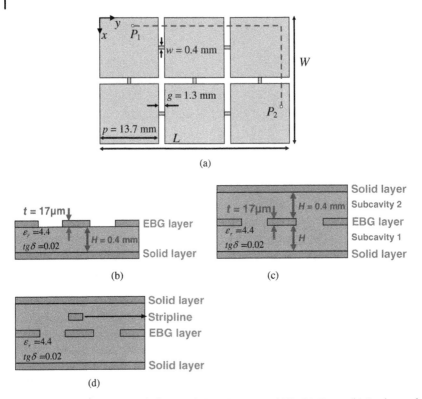

(a)

(b)

(c)

(d)

Figure 3.7 (a) EBG patterned plane with $L = 43.7$ mm and $W = 28.7$ mm. (b) Stack-up of a planar (not embedded) EBG. (c) Stack-up of an embedded EBG without signal lines (microstrip or stripline) [3]. (d) Stack-up of an embedded EBG with stripline in subcavity 2. (Taken from Ref. [3]. Reproduced with permission of IEEE.)

Figure 3.8. No bandgap is generated below 10 GHz. These resonances are due to the two smaller subcavities and also due to the larger one formed by the two external solid layers.

The resonances of the larger cavity are inhibited up to a given frequency and the bandgap is restored if the top and the bottom solid planes are electrically connected by means of stitching vias, placed in a regular pattern and crossing through the EBG layer by means of antipads [4]. This kind of configuration is named as embedded planar EBG. Figure 3.9a shows the intermediate patterned plane with the antipad holes, case with 41 stitching vias; Figure 3.9b shows the frequency spectrum of $|S_{21}|$ for the embedded planar EBG of Figure 3.7c with 25 and 41 regularly spaced stitching vias.

The embedded planar EBG with 41 shorting vias presents, in the bandgap, two peaks associated with the resonances of the cavities made by each single

Figure 3.8 Comparison between the $|S_{21}|$ for the two layers planar EBG geometry (Figure 3.7b, solid line) and the three-layer geometry (Figure 3.7c dashed line). (Taken from Ref. [3]. Reproduced with permission of IEEE.)

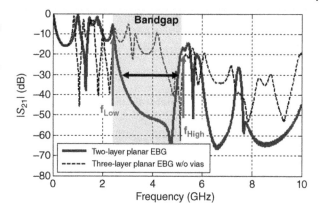

Figure 3.9 (a) EBG plane with antipads associated with the stitching vias from the top to the bottom solid plane, and (b) frequency spectrum of $|S_{21}|$ for embedded planar EBG with 25 (dashed line) and 41 (solid line) regularly placed shorting vias. (Taken from Ref. [3]. Reproduced with permission of IEEE.)

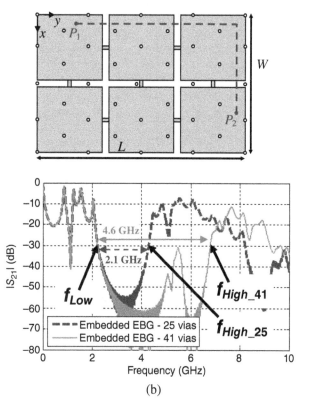

(b)

patch and its projection on the solid plane, whose amplitude is always below—30 dB. All the geometries investigated in this chapter are numerically simulated by means of CST Microwave Studio [1] using the finite integration technique (FIT) [5].

3.2.1 Striplines in a Symmetrical Embedded EBG Cavity

For a complete picture of the practical use of the embedded planar EBG, it is mandatory to analyze its impact on the signals transmission properties of traces routed in between the solid plane and the patterned layer as shown in Figure 3.7d. The signal integrity analysis of this structure is carried out adding a stripline, whose layout (in red dashed color) looking from the top is in Figure 3.7a between ports P_1 and P_2, in one of the two subcavities 1 or 2 (see Figure 3.7c). In both cases, the trace is located in the middle of the subcavity, symmetrically placed between the EBG plane and the solid one, as shown in Figure 3.7d. The stripline has a thickness $t = 17\,\mu m$ and width $w = 0.15\,mm$, giving nominal characteristic impedance (Z_c) of 50 Ω. The trace is terminated by two 50 Ω ports: P_1 (2 mm, 8 mm) and P_2 (20.7 mm, 41.7 mm) between the trace and the EBG layer. A third port, P_3 (5 mm, 5 mm), is added within the subcavity 1 between the solid and the patterned plane, in order to probe the coupling between the signal along the trace and the subcavity. This allows us to observe the mode conversion between the stripline mode and the parallel plane cavity mode [6].

Figure 3.10 shows three sets of curves: the transfer function of the embedded planar EBG (curve a), the transfer function $|S_{21}|$ of the stripline in the two

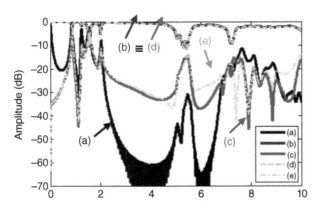

Figure 3.10 (a) Transfer function of the embedded planar EBG with 41 shorting vias. (b) $|S_{21}|$ of the stripline in subcavity 1. (c) $|S_{31}|$ with stripline in subcavity 1. (d) $|S_{21}|$ of the stripline in subcavity 2. (e) $|S_{31}|$ with stripline in subcavity 2. (Taken from Ref. [3]. Reproduced with permission of IEEE.)

Figure 3.11 E_z field inside the cavity at 3.5 GHz. The cut planes are identified by the dashed lines. (a) Side view. (b) Top view. (Taken from Ref. [3]. Reproduced with permission of IEEE.)

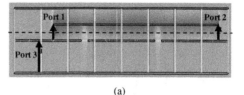

(a)

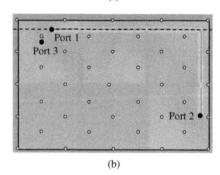

(b)

subcavities (curves b and d), and the stripline-to-subcavity 1 coupling when the stripline is placed in subcavity 1 or 2, respectively ($|S_{31}|$, curves c and e). The signal propagation along the stripline shows a flat behavior close to 0 dB, except at the peaks of the embedded planar EBG transfer function; at these frequencies, there are notches in the stripline $|S_{21}|$. Peaks in the cavity transfer function mean a high coupling into the cavity; thus, part of the electromagnetic energy associated with the stripline signal is diverted into the cavity; this leads to a reduction of the signal reaching the other stripline port. On the contrary, small values of the cavity transfer function mean a negligible coupling into the cavity; because of this, all of the signal electromagnetic energy is transferred and arrives unaffected at the other stripline port. A further validation of this result is given in Figure 3.11a and b where the spatial distribution of $|E_z|$ at 3.5 GHz (a frequency value into the bandgap) is shown when exciting Port 1. No field couples from the stripline to the subcavity 1 since no resonances are excited. This overall behavior reflects the results already obtained in Ref. [2] when considering a simple two-layer planar EBG structure. Very small differences are found in the stripline-to-subcavity coupling when moving the stripline from one subcavity to the other one.

Similar study is done investigating the SI for differential lines. The previous geometry is modified by only replacing the single-ended stripline with two differential traces ($w = 0.13$ mm) separated by $s = 0.25$ mm for having a nominal differential impedance $Z_c = 100\,\Omega$. Four separated ports are defined at the ends and are connected between the traces and the EBG plane. Figure 3.12 shows the power integrity performance of the embedded planar EBG structure compared

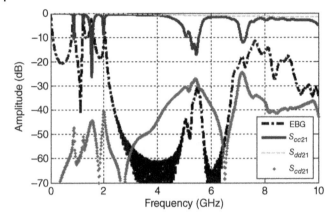

Figure 3.12 Transfer function of the embedded planar EBG compared to frequency spectrum of $|S_{dd21}|$, $|S_{cc21}|$, and $|S_{cd21}|$. (Taken from Ref. [3]. Reproduced with permission of IEEE.)

with the differential propagation scattering parameter $|S_{dd21}|$ and the common mode transfer function $|S_{cc21}|$, and also the mode conversion scattering parameter $|S_{cd21}|$. The frequency spectrum of $|S_{dd21}|$ has a flat response that is not affected by the patterned EBG plane, as also demonstrated in Ref. [2]. The $|S_{cc21}|$ has notches whenever resonant peaks occur within the cavity (same as the single-ended parameter). Also, the differential to common mode conversion S-parameter has peaks at the cavity resonances, because part of the input differential signal energy is converted into common mode at these frequencies. The behavior of the differential pair is exactly the same if the traces are routed within the subcavity 2 due to the perfect symmetry of the considered geometry.

3.2.2 Striplines in an Asymmetrical Embedded EBG Cavity

Most of the multilayer PCB layouts do not provide the geometric symmetry considered in the previous section because the dielectric layers have different thickness. This section focuses on the case where the two subcavities of the embedded planar EBG structure do not have the same thickness. A parametric analysis is made on the EBG configuration of Figure 3.9a (with the 41 shorting vias) by keeping the same distance $d = 0.817$ mm between the two solid planes, and changing the height h_1 of the EBG layer, as shown in Figure 3.13. The ports, at which will be computed $|S_{21}|$, are located in the subcavity 1, defined between the EBG layer and the bottom solid plane at the following coordinates P_1 (5 mm, 5 mm) and P_2 (38.7 mm, 23.7 mm), respectively. An equivalent low-frequency

Figure 3.13 (a) Stack-up of the embedded EBG for the parameterized power integrity analysis when moving the EBG layer. (b) Equivalent low-frequency circuit. (Taken from Ref. [3]. Reproduced with permission of IEEE.)

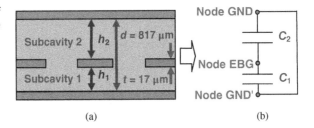

(a)　　　　　　　　　(b)

circuit is also derived. It consists of two capacitances C_1 and C_2 associated with the parallel plate subcavities 1 and 2. The two solid planes, identified by the nodes GND and GND', reduce at low frequency to the same node due to the 41 shorting vias. Therefore, C_1 and C_2 are in parallel, and the total capacitance between the EBG layer and the solid planes is $C_{tot} = C_1 + C_2$ where the two capacitances are computed as $C_i = \epsilon_0 \epsilon_r\, S/h_i$. $S = 0.0011$ m^2 is the effective EBG area and it is computed by adding the surface of the six patches to the surface of the seven bridges (the fringing field are neglected for simplicity) and h_i is the height of the ith subcavity. The parametric study consists in varying h_1 as $h_1 = x \cdot (d - t)$ with $x = 0.1, 0.3, 0.5, 0.7$, and 0.9.

Table 3.1 reports the values of C_1, C_2, and C_{tot} for the five cases. The five frequency spectra of $|S_{21}|$ are compared in Figure 3.14. In the below bandgap region (0–2 GHz), there are distributed resonances of subcavity 1 [2]. They are not the same for all the five cases because of C_{tot}: A large C_{tot} shifts downward the resonances. In this region, the curves (a) and (e) are overlapped, as well as the curves (b) and (d) due to the same values of C_{tot} associated with the capacitive circuit. In the bandgap region (2–6.5 GHz), the behavior is similar for the five cases. Above f_{High}, in the above bandgap region ($f > 6.5$ GHz), we observe a decreasing amplitude of $|S_{21}|$ from configuration (e) down to (a) as h_1 decreases: The thinner the cavity, the less energy coupled into it.

Table 3.1 Capacitance calculation.

x	h_1 (mm)	h_2 (mm)	C_1 (nF)	C_2 (nF)	C_{tot} (nF)
0.1	0.08	0.72	0.55	0.06	0.61
0.3	0.24	0.56	0.18	0.08	0.26
0.5	0.4	0.4	0.11	0.11	0.22
0.7	0.56	0.24	0.08	0.18	0.26
0.9	0.72	0.08	0.06	0.55	0.61

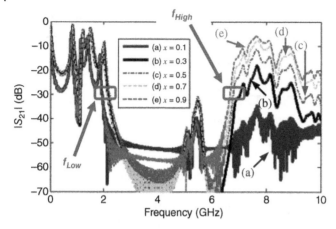

Figure 3.14 $|S_{21}|$ of the embedded planar EBG varying $h_1 = x \cdot (d - t)$. (Taken from Ref. [3]. Reproduced with permission of IEEE.)

A further analysis is done to investigate the impact of the variation of h_1 on the signal propagation along a single-ended stripline connected by the ports to the EBG layer and routed in the middle of the subcavity 1 of Figure 3.13. The trace thickness is $t = 0.017$ mm and its width is accordingly adjusted when changing h_1 for keeping the 50 Ω characteristic impedance. The simulation results are shown in Figure 3.15. The case (a) with $x = 0.1$ is

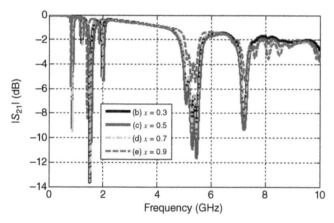

Figure 3.15 $|S_{21}|$ of the stripline in subcavity 1 varying $h_1 = x \cdot (d - t)$. (Taken from Ref. [3]. Reproduced with permission of IEEE.)

not considered because the trace width would have been too small, below any fabrication specifications. The trend of the four curves shows an expected trend: There are notches in the below bandgap region corresponding to the subcavity 1 resonances; there is a flat behavior in the bandgap region and the notches around 5.3 GHz and at 7.26 GHz are due to the coupling of the trace and the cavities formed by each single square patch and the two solid planes, respectively.

3.3 Conclusions

There is a strict correlation between the PI performances of an EBG structure and the SI performances of a trace routed over the patterned plane of such a structure. The chapter has studied the fundamental interactions associated with the EM coupling between the EBG cavity and the signal interconnects.

The design rules for planar EBG can be extended to embedded planar EBG [7–16] if a proper stitching of the two solid planes is performed. Peaks and notches of the embedded planar EBG cavity transfer function correspond to notches and high values of the single-ended trace transfer function. The $|S_{dd21}|$ is not significantly affected by the presence of the patterned plane, whereas $|S_{cc21}|$ behaves as $|S_{21}|$ of a single-ended interconnect.

References

1 Computer Simulation Technology, CST Studio Suite, 2016. Available at www .cst.com
2 F. de Paulis, A. Orlandi, L. Raimondo, and G. Antonini, Fundamental mechanisms of coupling between planar electromagnetic bandgap structures and interconnects in high-speed digital circuits. Part I: microstrip lines, in *Proc. of EMC Europe Workshop 2009, Athens, Greece*, June 11–12, 2009.
3 F. de Paulis, L. Raimondo, and A. Orlandi, Signal integrity analysis of embedded planar EBG structures, in *Proc. of the Asia-Pacific EMC 2010*, April 12–16, 2010, Beijing, China.
4 F. de Paulis, L. Raimondo, and A. Orlandi, Impact of shorting vias placement on embedded planar electromagnetic bandgap structures within multilayer printed circuit boards. *IEEE Trans. Microw. Theory Tech.*, Vol. 58, No. 7, 2010, pp. 1867–1876.

5 T. Weiland, A discretization method for the solution of Maxwell's equation for six-component fields. *Archiv Elektronik Uebertragungstechnik*, Vol. 31, 1977.

6 R. Rimolo-Donadio, X. Gu, Y.H. Kwark, M.B. Ritter, B. Archambeault, and F. de Paulis, et al., Physics-based via and trace models for efficient link simulation on multilayer structures up to 40 GHz. *IEEE Trans. Microw. Theory Tech.*, Vol. 57, No. 8, 2009, pp. 2072–2083.

7 B. Archambeault, *PCB Design for Real-World EMI Control*, Kluwer Academic Publishers, 2002.

8 M. Swaminathan and A.E. Engin, *Power Integrity Modeling and Design for Semiconductors and Systems*, Prentice Hall, Boston, MA, 2008.

9 T.L. Wu, Y.H. Lin, T.K. Wang, C.C. Wang, and S.T. Chen, Electromagnetic bandgap power/ground planes for wideband suppression of ground bounce noise and radiated emission in high-speed circuits. *IEEE Trans. Microw. Theory Tech.*, Vol. 53, No. 9, 2005, pp. 2935–2942.

10 B. Archambeault, F. de Paulis, A. Orlandi, and L. Raimondo, Common mode filtering performances of planar EBG structures, in *2009 IEEE International Symposium on Electromagnetic Compatibility (EMC2009)*, Austin, August 17–21, 2009.

11 D. Sievenpiper et al., High-impedance electromagnetic surfaces with a forbidden frequency band. *IEEE Trans. Microw. Theory Tech.*, Vol. 47, No. 11, 1999, pp. 2059–2073.

12 J. Choi, V. Govind, and M. Swaminathan, A novel electromagnetic bandgap (EBG) structure for mixed-signal system applications, in *Proc. of the 2004 IEEE Radio and Wireless Conference*, September 2004, pp. 243–246.

13 T. Kamgaing and O.M. Ramahi, A novel power plane with integrated simultaneous switching noise mitigation capability using high impedance surface. *IEEE Microw. Wirel. Compon. Lett.*, Vol. 13, No. 1, 2003, pp. 21–23.

14 S. Huh, M. Swaminathan, and F. Muradali, Design, modeling, and characterization of embedded electromagnetic band gap (EBG) structure, in *Proc of the 2008 IEEE Electrical Performance of Electronic Packaging Conference*, October 2008, pp. 83–86.

15 F. de Paulis, L. Raimondo, and A. Orlandi, IR-Drop analysis and thermal assessment of planar electromagnetic band-gap structures for power

integrity applications. *IEEE Trans. Adv. Packaging*, Vol. 33, No. 3, 2010, pp. 617–622.

16 F. de Paulis and A. Orlandi, Signal integrity analysis of single-ended and differential striplines in presence of EBG planar structures. *IEEE Microw. Wirel. Compon. Lett.*, Vol. 19, No. 9, 2009, pp. 554–556.

4

Planar Onboard EBG Filters for Common Mode Current Reduction

with contribution of Carlo Olivieri

This chapter will initially review the role of common mode (CM) filters in high-density PCB designs and the previously analyzed beneficial properties of the EBG structures. It will then introduce the concept of a CM filter based on patch resonant cavities and the operating principle underlying the construction of EBG-based CM filters. The main goal of the chapter is to introduce the design approach that has to be used in the building of EBG-based CM filters, and in particular the attention will be focused on the so-called onboard filters. As the name implies, these filters will include all the EBG structures directly implemented on the board stack-up. In Chapter 6, we will introduce EBG filters called "removable" that are separate part and mounted on top of the PCB.

The design workflow will be described in detail so that the readers will be able to understand which are the main design steps and the related issues. In this chapter, different design strategies will be described and proper considerations will be stated in order to target the EBG filter to specific design purposes, such as the improvement of the filter attenuation and bandwidth, or to minimize the occupied surface. As a concluding section, the main advantages and disadvantages of the considered structures will be described so that the optimum design solution can be implemented.

4.1 EBG Structures as Common Mode Filters: Overview and Operating Principles

Differential signaling is a common layout practice used to design high-speed digital buses in printed circuit boards (PCBs) and packages. Ideally, if perfectly implemented, it reduces the amount of common mode noise seen at the receiver and also the radiated emissions from the CM current and thus

Electromagnetic Bandgap (EBG) Structures: Common Mode Filters for High-Speed Digital Systems,
First Edition. Antonio Orlandi, Bruce Archambeault, Francesco De Paulis, and Samuel Connor.
© 2017 by The Institute of Electrical and Electronics Eingineers, Inc. Published 2017 by John Wiley & Sons, Inc.

electromagnetic interference (EMI) and cross talk problems, and it increases the noise immunity from the surrounding environment allowing higher data rates. In the real world, small asymmetries in the differential pair can have a large impact on the common mode magnitude and can give rise to electromagnetic compatibility (EMC) and signal integrity (SI) issues if not handled properly [1]. These asymmetries can be caused by both geometrical properties, for example, different trace length, and electrical properties, for example, imbalances in the two transmitted signals such as rise/fall time mismatch, time delay (skew), or even amplitude mismatch between the two single-ended signals [2]. The main effect is the conversion of some of the differential mode (DM) signal to common mode, which, in turn may be responsible for electromagnetic interference emissions problems.

In a real-world design, completely removing the asymmetries is impractical or impossible; thus, a suitable solution to reduce the CM harmonic components is to filter the CM portion of the signal. This filtering operation is usually achieved through the use of discrete components, which have several disadvantages: They take up space on the board, provide an additional cost, and are often the cause of undesirable attenuation of the intentional differential signal.

As discussed in Chapter 2, the planar electromagnetic bandgap (EBG) structures are periodic structures that consist of two planes: one is a solid plane and the other is a patterned plane suitably shaped in order to obtain a high-impedance surface, which has the property of inhibiting the electromagnetic field propagation within a certain range of frequencies (the bandgap) [3]. They were introduced for antenna applications, but their use has been extended to PCBs and packages, mainly applied to power integrity (PI) issues [4–7]. Recently, their impact on the signal integrity of digital signals has been considered [8–10]. The EBG patterned plane can be seen as the periodic repetition of unit cells; each unit cell consists of a patch (square or rectangular) connected to the adjacent one through a bridge, which can assume several shapes [11–14]. In Refs [9,15], it has been demonstrated that there is a strict correlation between the PI performance of an EBG structure and the SI performance of a single-ended trace routed over the patterned plane of such a structure. In particular, each one of the resonance peaks in the PI transfer function between the two excitation ports corresponds to a notch in the SI transfer function between the signal trace ending ports. Thus, by designing the patterned plane of the EBG structure in a suitable way, it is possible to selectively attenuate the undesired harmonics of the signal. Furthermore, the filtering performance from a single-ended trace can be extended to the CM harmonic components of a differential pair of traces [15,16].

Since the notches of the SI transfer function are related to the peaks of resonance of the cavity made by the single patch and its projection on the solid plane underneath, the original EBG structure is modified by removing the bridges that connect two adjacent patches. In this way, each one of the single patch subcavities is excited when the traces cross the gap between two adjacent patches. This kind of patch isolation allows one to analytically predict the resonant frequencies of the patch itself and to design it to filter out the undesired CM harmonics [17].

The origins of the common mode signals will be illustrated and then the basic EM operating principle underlying the building of proper EBG-based filtering structures will be introduced; this will constitute the preparatory matter for the rest of the chapter.

4.1.1 Origins of Common Mode Signals and Noise

Nowadays, since data rates are approaching several hundreds of megabits or gigabits per second (Mb/s or Gb/s), signal integrity concerns will require that differential signaling is properly used in order to ensure the required signal quality. Dielectric loss for long traces, reflections from connectors and vias, and even surface roughness will reduce signal quality at the end of long traces at very high data rates. Differential signaling is also more immune to external noise corrupting the intentional signals. The basic intention for differential signals is for equal and opposite currents (and voltages) to exist on the pair of traces, and the ground-reference plane plays no role in the intentional signal current. In reality, this is true only when there are only two signal conductors in free space, with no other metal nearby. This perfect condition never occurs in typical printed circuit boards.

For the sake of completeness, some details will be given regarding the generation of detrimental common mode signals, in particular focusing on the most frequently observed unbalance occurring in PCB design: presence of a certain amount of skew, rise and fall time mismatches, and differential signals amplitude mismatch [1,2].

4.1.1.1 Effects of Skew

In order to analyze such a kind of unbalance, an example pseudodifferential signal of 2 Gb/s (500 ps pulse width) square wave with a nominal rise/fall time of 50 ps was considered. In particular, the second output of the complementary driver was delayed, introducing skew from 10 to 200 ps. This delay represents a skew of 2–40% of the pulse width. The resulting individual channel voltage waveforms are shown in Figure 4.1, whereas the resulting differential pulse at the receiver is created by subtracting the two signals, and is shown in Figure 4.2. Note that while the differential pulse shows some distortion from the skew of the two original complementary single-ended signals, the final differential pulse is acceptable even with large amounts of skew in the 10–20% of the bit width. The amount of apparent distortion is reduced when the rise/fall times are slowed slightly.

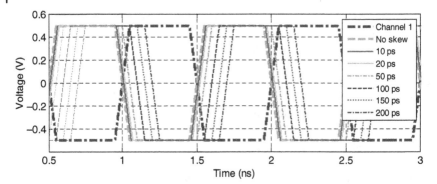

Figure 4.1 Individual channels of differential signal with in-pair skew 2 Gb/s with 50 ps rise/fall time (+/−1 V). (Reproduced with permission from Ref. [2].)

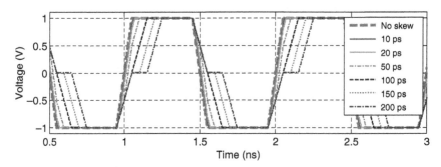

Figure 4.2 Differential signal with in-pair skew, 2 Gb/s with 50 ps rise/fall time (+/−1 V). (Reproduced with permission from Ref. [2].)

For EMC applications, the common mode signal is more important. Figure 4.3 shows the common mode voltage for the various amounts of skew. The common mode voltage was found by adding the two signals for every instant in time. As the skew increases, the amplitude of the common mode voltage increases. Note that once 100 ps of skew is reached, the amplitude of the common mode signal assumes a peaked waveform. Figure 4.4 shows the frequency domain harmonics for the common mode voltage. Only odd harmonics are created since the rise/fall times are identical for this square wave example.

4.1.1.2 Effects of Rise/Fall Time Mismatch

The previous section showed how rapidly the skew of the pseudodifferential signals can cause common mode voltage. This section examines instead the effect of rise time and fall time mismatch between the two complementary signals. Figure 4.5 shows the differential signals when the rise/fall times are varied from 50 to 200 ps. The differential signal looks good, even for the extreme

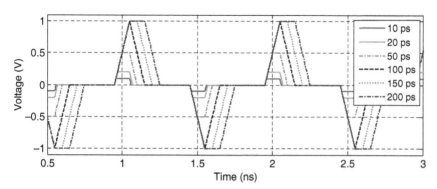

Figure 4.3 Common mode voltage on differential pair due to in-pair skew, 2 Gb/s with 50 ps rise/fall time (+/−1 V). (Reproduced with permission from Ref. [2].)

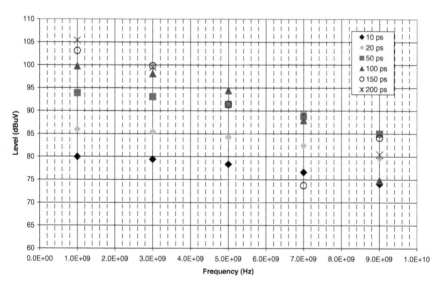

Figure 4.4 Harmonic content of the common mode signal due to in-pair skew. (Reproduced with permission from Ref. [2].)

case of 50/200 ps rise/fall time. However, Figure 4.6 shows the individual channels and the resulting common mode; attention must be paid to the fact that there is a pulse at both the rising and falling edges, indicating that the harmonic frequencies will be even harmonics of the fundamental frequency of the differential signal.

The resulting common mode voltage is shown in Figure 4.7 and the harmonics are shown in Figure 4.8. In all the considered cases, there can be significant increases of harmonics amplitudes with respect to the case when

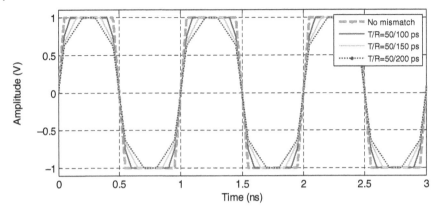

Figure 4.5 Differential mode voltage on differential pair due to rise/fall time mismatch, 2 Gb/s with differential signal +/−1 V.

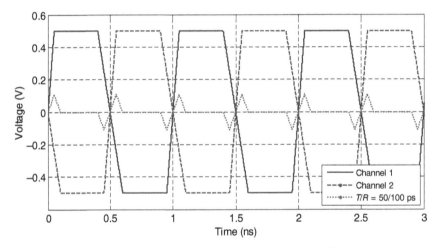

Figure 4.6 Example of effects on common mode signal when the differential signals are affected by rise/fall time mismatch, 2 Gb/s square wave (rise/fall = 50 and 100 ps).

there is no mismatch. Nevertheless, the case in Fig. 4.8 where the rise is 50 ps and fall is 55 ps is quite close to the case with no mismatch. Clearly, no mismatch would result in zero CM voltage. Significant common mode voltage was created with rise/fall time mismatch. Even though this CM signals are not as large as the common mode signals caused by skew, the real-world signals will combine both the causes (skew and rise/fall time threshold); thus, the resulting total common mode noise signal will be a challenging source of EMI emissions.

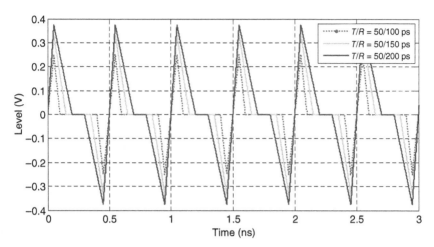

Figure 4.7 Common mode voltage due to different amounts of rise/fall time mismatch, 2 Gb/s with differential signal +/−1 V.

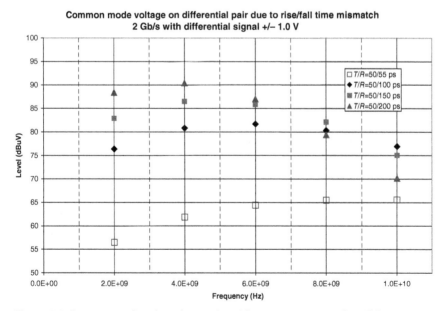

Figure 4.8 Common mode voltage harmonics with varying amounts of rise/fall time mismatch. (Reproduced with permission from Ref. [2].)

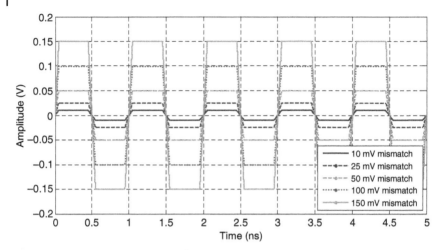

Figure 4.9 Common mode voltage on differential pair due to various amounts of amplitude mismatch, 2 Gb/s nominal differential signal with 100 ps rise/fall time +/−1 V.

4.1.1.3 Effects of Amplitude Mismatch

Amplitude mismatch between the two individual channels can also have a significant impact on common mode signals. Figure 4.9 shows the time domain of the common mode signals created with small amounts of amplitude mismatch. The differential signal is a +/− 1 V square wave with a rise/fall time of 100 ps. The amplitude mismatch was created using individual channels with a 1 V (peak–peak) square wave, and the mismatch added to one channel only. The harmonic amplitudes characterizing the common mode signals created from the amplitude mismatches are shown in Figure 4.10. Once again, only the odd harmonics are created since the rise/fall times of the square wave are identical. The amplitude of each harmonic grows the same (even though the starting amplitude is different). Figure 4.10 shows the harmonic amplitude growth and it can be seen that small percentages of amplitude mismatch can create a significant common mode signal amplitude.

4.1.1.4 Differential Vias

Since multilayer PCBs are commonly used today, when differential signals are transitioning from one layer in a board to another layer and changing reference layers, there is an opportunity for common mode noise to be created. This differential-to-common mode conversion occurs when there is any asymmetry for the differential vias. As an example, Figure 4.11 shows a differential pair of vias with signals transitioning from the top layer to the bottom layer, with a 10 mil spacing between the two reference (or "ground, GND") planes. A single via that connects between the two reference planes is moved from close to one of

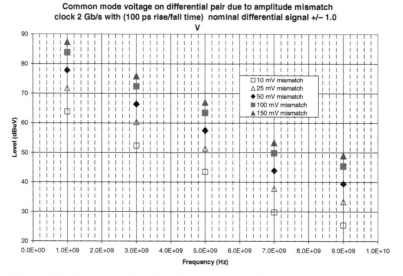

Figure 4.10 Common mode voltage harmonics when there are various amounts of amplitude mismatch on the differential pair signals. (Reproduced with permission from Ref. [2].)

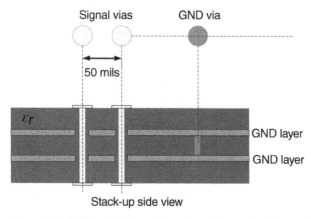

Figure 4.11 Differential vias with nearby reference (or "ground, GND") via.

the differential vias to further distances and the amount of mode conversion is determined. Figure 4.12 shows the differential-to-common mode conversion when the via runs through a number of layers. Each layer transition adds more mode conversion and higher amplitude of common mode noise on the differential pair. While the mode conversion is relatively low, the distance to the reference (or "ground, GND") via can make a 40 dB difference.

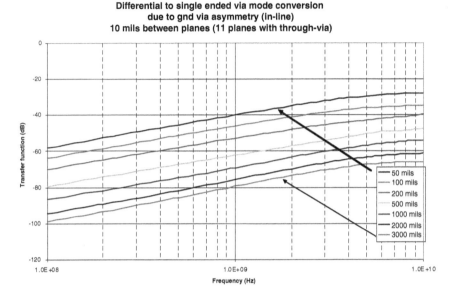

Figure 4.12 Differential mode-to-common mode conversion through 11 layers due to GND via distance. (Reproduced with permission from Ref. [2].)

4.1.2 Resonant Patch-Based Common Mode Filters: EM Operating Principle

After providing the overview of the common mode noise generation, the proper countermeasure needs to be addressed; the attention will be focused only on CM filters developed employing resonant cavities and EBG structures.

In particular, in order to explain the basic concept and the operating principle of an EBG-based CM filter, let us first take into account the basic layout depicted in Figure 4.13; it refers specifically to an EBG common mode filter for differential pairs in multilayer PCBs. It must be noted that it is essentially a filter based on resonant patch cavities and hence the filtering action is based on the coupling of electromagnetic (EM) energy between the common mode return current and the cavity made by each metal patch (that acts as reference/return conductor for the differential pair) and the solid plane underneath. This coupling is achieved at the gap location and at the cavity resonant frequencies. As described in the previous chapter, the basic structure in Figure 4.13 is characterized by some fundamental geometrical parameters. For our purposes, in the rest of this chapter, the main parameters will be the following—a: patch width, g: gap between patches, S_m: distance of the traces from the upper edge of the patches, d: thickness of the dielectric layers, W_m: width of the differential microstrips, S: separation between the traces. Figure 4.14 provides a detailed explanation of the basic phenomena underlying the aforementioned EM

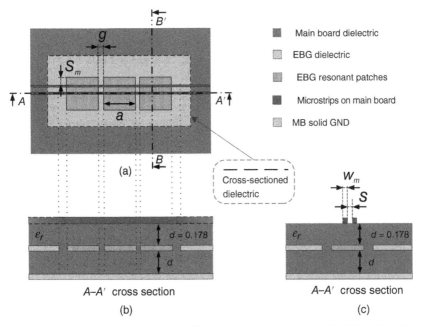

Figure 4.13 EBG-based common mode filter layout and stack-up details. (a) Top view. (b and c) Cross-sectional side views.

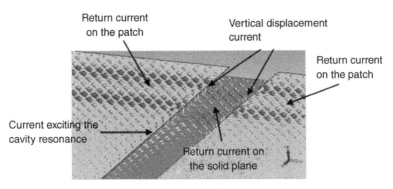

Figure 4.14 Return path of the common mode return current.

coupling, especially looking at the return current on the patch: The return current that reaches the gap sees the gap discontinuity and it is forced to flow down to the solid plane, and then up again on the top surface of the following patch. The displacement current flowing from the patch to the solid plane is normal to the plane surface; thus, it can excite the cavity at its resonant

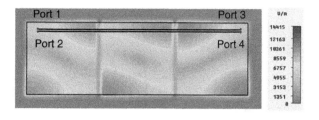

Figure 4.15 Resonance of the patch cavity at 5 GHz (TM_{01}/TM_{10} modes): pattern of the amplitude of the component of the electric field E normal to the plane of the patches.

frequencies. The first resonance of one of the three square cavities in Figure 4.13 occurs at 5 GHz due to the specific dimensions: $a = 15.6$ mm, $g = 1.3$ mm, $S_m = 0.25$ mm, $S = 0.25$ mm, $W_m = 0.25$ mm, and stack-up data (it is related to the TM_{10} and TM_{01} modes).

In order to better understand this basic EM coupling underlying the CM filter functioning, the spatial distribution of the E field is computed by using a finite integration technique full-wave three-dimensional code in time domain [18].

The normal component to the plane of the patches of the electric (E) field at 5 GHz is shown in Figure 4.15, highlighting the resonant behavior of the cavity that combines both the modes occurring at this frequency.

This is the basic principle that will enable the manufacturing of simple resonant patch-based CM filters and subsequently, in turn, the design of more elaborated EBG-based CM filters.

4.2 Resonant Patch-Based Common Mode Filters: Basic Behavior and Features

4.2.1 Basic Behavior of CM Filters Based on Resonant Patches

The fundamental mechanisms and the relevant properties of the CM filters based on resonant patches are discussed in this section. Figure 4.16 shows the top view and the stack-up of such a CM filter together with its geometrical dimensions. The differential pair is routed on top of the patterned plane containing all the patch structures; thus, the input voltage ports related to a microstrip pair are referenced to this plane. The filtering effect is achieved from the resonant behavior of the cavity made by each single patch and the solid reference plane underneath. Therefore, as a consequence, in practical applications of a real PCB layout, no vias should be placed within the patch area and no connection should be made between the patch and the reference plane. In this particular case, the filter layout can also be thought as merely based on a planar EBG structure (as introduced in Chapter 3) whose connection bridges have been removed.

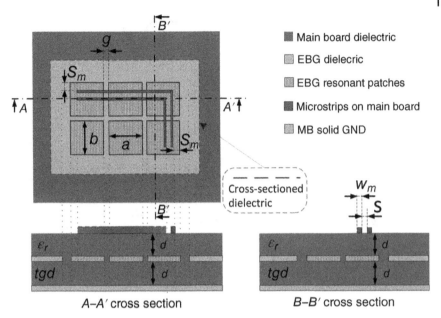

Figure 4.16 Top view and stack-up of a CM filter based on a planar EBG structure.

The transfer function of the resonant cavity without bridges, named $|S_{21}^{PI}|$, together with the common and differential mode components of the signal, $|S_{cc21}^{SI}|$ and $|S_{dd21}^{SI}|$, are illustrated in Figure 4.17. Two aspects must be highlighted: the first is related to the correspondence between the peaks of S_{21}^{PI} and the notches of the S_{cc21}^{SI}; the second one is that the S_{dd21}^{SI} is not affected by the presence of the patterned plane. Figure 4.17 also shows the transfer parameter (S_{21}^{SI}) of a single-ended microstrip laid out in between the differential trace pair; its trend is very similar to the $|S_{cc21}^{SI}|$, as expected, since the current propagation and the trace to cavity coupling mechanisms are fundamentally the same in both cases. Therefore, this simplified single-ended geometry can be employed for a faster analysis of the common mode propagation under interest, according to what already introduced in Chapter 3.

The frequency of the peaks in $|S_{21}^{PI}|$ (and the notches in $|S_{cc21}^{SI}|$) are related to the single patch cavity and can be predicted by applying Equation 4.1, which comes from the theory of planar resonators [19], as also discussed in Chapter 2:

$$f_{TM_{mn}} = \frac{1}{2\pi\sqrt{\mu\varepsilon}} \sqrt{\left(\frac{m\pi}{a}\right)^2 + \left(\frac{n\pi}{b}\right)^2} \tag{4.1}$$

In (4.1), a and b are the dimensions of the patch, as shown in Figure 4.16. The results illustrated in Figure 4.17 come from a planar structure with a square

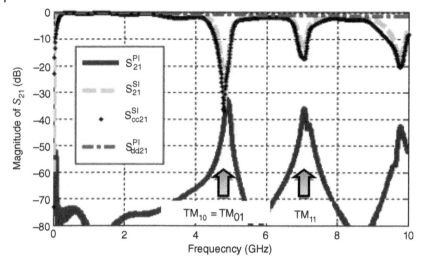

Figure 4.17 Correspondence among peaks and notches of $|S_{21}^{Pl}|$ (solid) and $|S_{cc21}^{Sl}|$ (dashed). (Reproduced with permission from Ref. [2].)

patch, so (4.1) can be applied to the first TM_z modes of the cavities in Figure 4.16 to calculate the first three resonance modes, indicated by the arrows in Figure 4.17. The expressions for the $TM_{z_{10(01)}}$ and the $TM_{z_{11}}$ modes of a square patch are given in (4.2):

$$\begin{cases} f_{TM_{10}} = f_{TM_{01}} = \dfrac{c_0}{2a\sqrt{\varepsilon_r}} \\[2ex] f_{TM_{11}} = \dfrac{c_0\sqrt{2}}{2a\sqrt{\varepsilon_r}} \end{cases} \tag{4.2}$$

Thus, it is possible to determine the dimensions of the patches employed by the CM filter in a simple way, that is, starting from the frequency of the harmonic component that must be filtered out.

The relationships presented until now can be retained useful for a rough preliminary sizing of the filter geometry. More accurate details on a general design workflow will be given later in this chapter, in particular including the case where instead of simple patches, we have EBGs (that is to say, patches connected by bridges).

For the specific example geometry shown in Figure 4.16, the filter target frequency can be set, for example, to 5 GHz according to common application

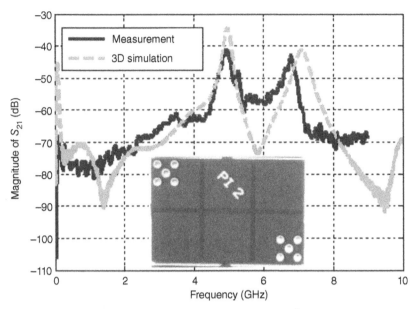

Figure 4.18 Comparison between measured and computed S_{21}^{PI}; merged in the plot there is also the photograph of the real board under measurement.

requirements. The starting geometrical details come from the employed stack-up data: $\varepsilon_r = 3.7$, $tg\delta = 0.02$, $d = 0.178$ mm, and from the design of the differential traces, accounting for $S_m = 0.25$ mm, $S = 0.25$ mm, and $W_m = 0.25$ mm. Applying the aforementioned preliminary design procedure, the following dimensions have been determined for the resonant patch cavity: $a = 15.6$ mm, $b = 15.6$ mm, $g = 1.3$ mm, which will provide the desired filtering action. The presented approach for the CM filter design is also further validated through a preliminary model to hardware correlation analysis. A simple two-layer printed circuit board is built for analyzing the power integrity behavior of the patterned cavity. The terminals of two SMA connectors are connected to the patches (the central pins) and to the solid plane (the return pins). The comparison between the measured and the simulated results are provided in Figure 4.18, in which the S_{21}^{PI} transfer function have been reported. The main feature of the structure, the resonance at 5 GHz, is achieved, confirming the correct resonant behavior of such layout technique. The second resonance should be at 7.07 GHz (as from the simulated data), but the FR-4 dielectric used in the board, with poor characteristics at high frequency, does not allow a correct match of this second peak (it occurs in fact at 6.8 GHz). This illustrated the need to know the correct characteristics of the dielectric used.

4.2.2 Basic Features of Resonant Patches CM Filters

4.2.2.1 Effect of Patch Number and Trace Layout

As one would expect, when a differential pair crosses more than one gap, the energy conversion from the common mode to the cavity mode will be larger, and thus the excitation of the patch cavities adjacent to the crossed gaps is increased. Since the gained notch depth in the common mode transfer function (S_{cc21}) is a direct consequence of this energy coupling, it results that more crossed gaps (patches) result in a deeper notch. A proper analysis for the quantification of the impact originating from the number of gap crossings is done simulating a single-ended microstrip instead of the differential pair. It has been shown [15,16] that there is duality between S_{21} of an equivalent single-ended trace and S_{cc21} of a differential pair; thus, single-ended models are used for the next simulations introduced in this section. A 50 Ω single-ended microstrip trace is designed based on the stack-up in Figure 4.16. The analysis is carried out on four models with different number N_p of patches (N_p from 3 to 6); thus, the microstrip crosses N_c gaps (N_c from 2 to 5), as depicted in Figure 4.19. Furthermore, the trace is moved along the gap from the patch edge varying S_m, with $S_m = 2, 3.93, 5.86$, and 7.8 mm, respectively. The latter case has the trace located at the patch center.

As previously done, the patch size is designed to achieve a first resonant mode at 5 GHz ($a = 15.6$ mm); the simulated structures present the same notch frequency at around 4.94 GHz, thus giving an error on the design frequency less than 2%. All the plots in Figure 4.20 show the same trend, validating the assumption on the coupling mechanisms discussed above; as more gaps are crossed by the trace, the S_{21} notch is deeper. A further deduction can be drawn comparing the four plots. The notch depth is also function of the trace position, as can be clearly seen comparing Figure 4.20a–d. The maximum depth

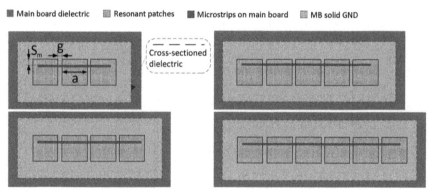

■ Main board dielectric　▓ Resonant patches　■ Microstrips on main board　▓ MB solid GND

Figure 4.19 Geometry of the models considered for the analysis of different patch crossings (N_p from 3 to 6).

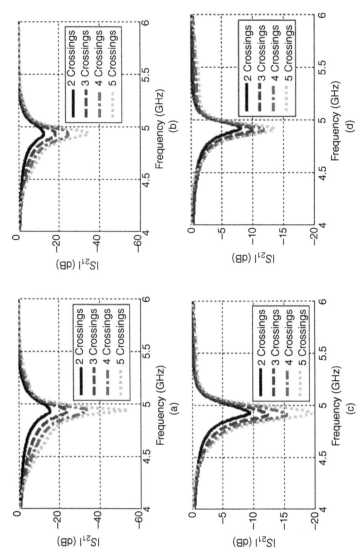

Figure 4.20 S_{21} of a trace crossing two to five gaps at various trace to patch edge distances. (a) $s_m = 2$ mm. (b) $s_m = 3.93$ mm. (c) $s_m = 5.86$ mm. (d) $s_m = 7.8$ mm. (Reproduced from Ref. [20] with permission of IEEE.)

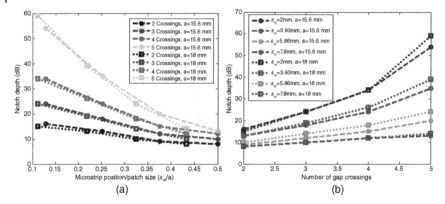

Figure 4.21 Notch depth analysis. (a) As function of the microstrip position. (b) As function of the number of gap crossings N_c. (Reproduced from Ref. [20] with permission of IEEE.)

(associated with the five crossing case) is 54, 33, 20, and 13 dB, when moving the trace from the edge ($S_m = 2$ mm) to closer to the patch center ($S_m = 7.8$ mm). The explanation of this phenomenon is related to the E-field distribution associated with the first resonant mode ($TM_{01/10}$). The field distribution is given by the combination of the TM_{10} and the TM_{01} mode, as shown in Figure 4.15, and the notch depth is maximum closer to the lateral gap end.

Therefore, more energy can be coupled from the common mode propagating along the trace pair to the cavity mode at this specific resonance, providing a deeper notch.

Additional simulations are done for a further validation of the results regarding the notch depth. The patch size is changed from $a = 15.6$ mm to $a = 18$ mm. This latter dimension leads to a resonant frequency of 4.3 GHz (instead of the previous 5 GHz). The dependence of the notch depth to N_c follows the same trend shown in Figure 4.20. Figure 4.21a shows the notch depth as function of the trace position; the curve trend, similar for the two patch models ($a = 15.6$ mm and $a = 18$ mm) with different number of crossings, confirms the assumption of greater coupling when the trace is closer to the patch edge. Figure 4.21b demonstrates the impact of the number of gap crossings on the notch depth.

4.2.2.2 Multiple Crossings on the Same Gap
Several problems could occur during a real PCB layout; one of the most common among them is related to the surface area that must be reserved for the common mode planar filter design (also when using EBG resonating structures instead of simple metal patches). Even though the theoretical study carried out in the previous section suggests to increase the number of crossings for obtaining a deeper notch, the available layout area might not be adequate to achieve the notch depth from the required design specifications. It is a matter of

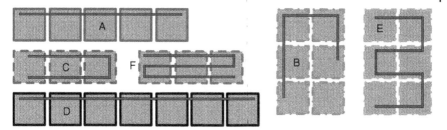

Figure 4.22 Layout configurations for studying the multiple crossing technique.

searching for a good trade-off between occupied area and gained notch depth. A technique to bypass this possible limitation is now presented and it is based on the idea of exploiting multiple crossings of the same gap by the differential pair of interest.

Several models were simulated for investigating this alternative layout strategy. These layout configurations are shown in Figure 4.22. The first three cases (A, B, and C) include four crossings, whereas the other three models (D, E, and F) have six crossings. These geometries cover different layout topologies: array of patches (A and D), a 3×2 matrix (B and E), and 1×3 array (C and F).

The simulation results related to these six models are shown in Figure 4.23. In particular, Figure 4.23a is related to the cases with four gap crossings, whereas Figure 4.23b shows the results from the six gap crossings cases. First of all, looking just at the first resonant mode at about 5 GHz, the six crossing cases provide a deeper notch. The cases A and B reach the same depth, about -32 dB. Case C, instead, has a -26 dB notch at 5 GHz. This reduction is due to the multiple crossing layout (both gaps are crossed twice). This conclusion can be stated also looking at the other cases related to the six crossing configurations. Case D does not have any repeated crossing (1×7 straight array of patches) and it provides the larger notch depth, -56 dB. The model E has one multiple crossing and it experiences a notch reduction of 4 dB (-52 dB). The last model F has the two gaps crossed three times, making the notch depth smaller, at -33 dB.

Table 4.1 summarizes the performances of the studied layout configurations providing the relationships between the number of independent/repeated crossings and the depth of the notch related to the first patch resonance. From the results it appears clear that the independent crossings are more effective (the generated notch is more deep) than the repeated one. The latter infact are subjected to the mutual crosstalk that decrease the depth of the associated notch.

4.3 Resonant Patch-Based Filters: Experimental Validation

In order to experimentally validate the proposed CM filtering technique, a 16-layer test board has been designed for applying the resonant-based CM filters

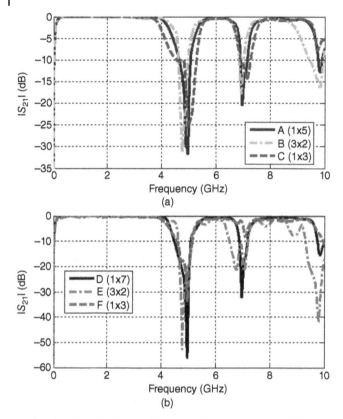

Figure 4.23 $|S_{21}|$ for the models with (a) four crossings and (b) six crossings. (Reproduced from Ref. [20] with permission of IEEE.)

Table 4.1 Simulation results: relationships between the number of independent/repeated crossings and the depth of the notch.

Model	Number of independent crossings	Number of repeated crossings	Notch depth (dB)
A	4	0	−32
B	4	0	−31
C	0	4	−26
D	6	0	−56
E	4	2	−52
F	0	6	−33

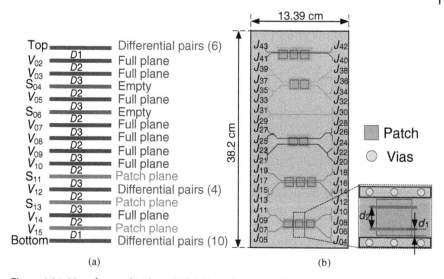

(a)   (b)

Figure 4.24 Manufactured 16-layer PCB. (a) Board stack-up. (b) Combined top view of the EBG layer V_{15} (same layout of layers S_{11} and S_{13}) and of the bottom layer with 10 pairs of differential microstrips.

described in the previous section. For this study, the filters are designed to suppress the first common mode harmonic of a 10 Gbps differential signal corresponding to a fundamental frequency of 5 GHz. The basic unit cell is a square patch dimensioned to have its first resonant mode ($TM_{10} = TM_{01}$) exactly at 5 GHz. The board stack-up is given in Figure 4.24a. The metal thickness is 35 μm, and the dielectric heights are $D1 = 0.142$ mm, $D2 = 0.127$ mm, and $D3 = 0.132$ mm. The nominal relative dielectric constant is $\varepsilon_r = 3.7$ and the loss tangent is $\tan \delta = 0.04$. Figure 4.24b shows the layout of the 10 differential microstrip pairs on the bottom layer and the corresponding EBG patches on layer V_{15}. The layout of V_{15} is equal to that of layers S_{11} and S_{13}. The layout of the EBG consists of square patches separated by the rest of the metal plane by a gap. The differential signal pairs are partially routed on top of the patch area and partially on top of the solid plane on the same layer. Some filters are implemented on the external layer V_{15} and some other deeper in the stack-up (layers S_{11} and S_{13}). The size of each square patch is $a = 14$ mm, the gap at layers S_{11}, S_{13}, and V_{15} among adjacent patches and between the patches, and the solid plane is $g = 1.27$ mm. Molex SMA connectors, attached to the trace ends for performing the S-parameter measurements, are identified in Figure 4.24b by letters J_{04}–J_{43}.

All the differential traces on the bottom layer are 168 μm wide and have 192 μm edge-to-edge separation. They run next to each other on top of the patches and then diverge on the solid plane for allowing the connector placement. The distance between the closest edge of the patch and the center of the microstrip pair is identified as d_1. The parameter d_2 is the distance between two microstrips, when they are placed within the same patch, as shown in Figure 4.24b.

4.3.1 Model to Hardware Correlation

The filtering impact on the differential pairs given in Figure 4.24b is measured and numerically simulated for validating the layout technique proposed in the previous chapter for a patch-resonant common mode filter. The measured microstrip pairs are those identified by the connectors *J28–J30* and *J40–J42*. The value d_1 is 3.4 mm for the *J40–J42* case. The case *J28–J30* is considered as the reference since the traces are laid out on a solid area of the layer V_{15}. Through-hole vias connecting layers V_{14} to V_{15} surrounding the filter areas are added in the layout to avoid the excitation of unwanted resonances associated with the cavity formed by the full layers V_{15} and V_{14}.

Some details on the via position are given in the inset of Figure 4.24b; the via–via distance is 2.54 mm, the via diameter is 0.4 mm, and the via–gap distance is 1.5 mm. The only filter without vias is the one associated with the *J40–J42* differential pair. The comparison between simulated and measured results is shown in Figure 4.25a for the reference case (no patch structures) *J28–J30*. The computed and measured magnitude S_{dd21} and S_{cc21} parameters show clearly lossy behavior without any resonances [10]. The difference between simulated and measured data is within 4 dB up to 15 GHz. The increasing dielectric losses associated with the manufactured board make the comparison worse beyond 15 GHz. In order to capture this effect, a first-order Debye model for the dielectric is set in the simulation in order to have the nominal value of $\varepsilon_r = 3.7$ and a maximum value of tan$\delta = 0.04$ at 10 GHz, which is the center frequency of the considered frequency range. According to the IEEE Standard P1597 [21], Figure 4.25b and c shows the Global Difference Measure (GDMc), the figure of merit of the Feature Selective Validation (FSV) [22,23] technique that quantifies the comparison between the measured and the simulated data, for $|S_{dd21}|$ (Figure 4.25b) and $|S_{cc21}|$ (Figure 4.25c), respectively.

Figure 4.26a shows the comparison between measured and simulated results for the case *J40–J42*. The $|S_{dd21}|$ is not significantly affected by the nonideal reference, as mentioned in Ref. [10]. Also, its magnitude is comparable with that of case *J28–J30* in Figure 4.25a, for both the measured and the simulated data. The measured $|S_{cc21}|$ shows a notch at 5.16 GHz, whereas the simulated frequency spectrum data have the same notch at 5.3 GHz. Figure 4.26b and c reports the values of GDMc for the comparisons of measured and simulated frequency spectra of $|S_{dd21}|$ and $|S_{cc21}|$ shown in Figure 4.26a, respectively. The 2.7% difference between the simulated and measured notch frequency allows

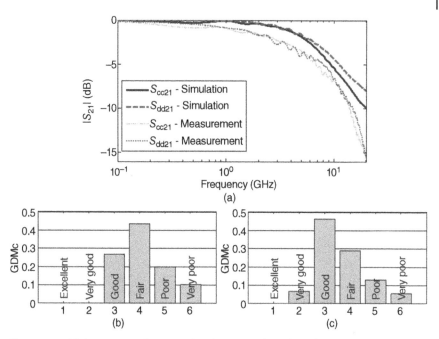

Figure 4.25 (a) Comparison of measured and simulated $|S_{dd21}|$ and $|S_{cc21}|$ for the case J_{28}–J_{30}. (b) GDMc for the comparison of measured and simulated $|S_{dd21}|$ in part (a) (GREAD = 5, SPREAD = 3). (c) GDMc for the comparison of measured and simulated $|S_{cc21}|$ in part (a) (GREAD = 5, SPREAD = 3). (Reproduced from Ref. [16] with permission of IEEE.)

one to consider the proposed layout design technique reliable for a pre-layout design step.

4.3.2 From Frequency Domain Measurements to Time Domain Simulations

The experimental S-parameter data measured as described in the previous section are used herein for running a simulation in the time domain. This test helps understanding the effects of the common mode filter on the common mode noise associated with a clock signal. A 5 GHz differential clock is considered in this analysis with 1 V peak–peak amplitude. The swing of two single-ended signals goes from −0.5 to 0.5 V. The rise time, same as the fall time, is 30 ps. A 15 ps delay is artificially introduced between the two single-ended signals to generate an input common mode noise, as shown in Figure 4.27. The spectrum of the common mode signal is evaluated through the fast Fourier transform (FFT), and it is shown in Figure 4.28 together with the $|S_{cc21}|$ of the two differential pairs $J28$–$J30$ and $J40$–$J42$ shown in Figure 4.25 and Figure 4.26, respectively. It is clear that the first harmonic

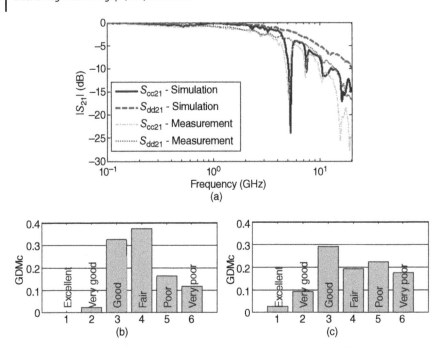

Figure 4.26 (a) Comparison of measured and simulated $|S_{dd21}|$ and $|S_{cc21}|$ for the case J_{40}–J_{42}. (b) GDMc for the comparison of measured and simulated $|S_{dd21}|$ in part (a) (GREAD = 5, SPREAD = 3). (c) GDMc for the comparison of measured and simulated $|S_{cc21}|$ in part (a) (GREAD = 6, SPREAD = 4). (Reproduced from Ref. [16] with permission of IEEE.)

of the common mode signals corresponds exactly to the notch of the $|S_{cc21}|$ associated with the $J40$–$J42$ differential trace. $|S_{cc21}\_J_{40-42}| = -8.4$ dB, whereas $|S_{cc21}\_J_{28-30}| = -1.8$ dB; thus, the first common mode harmonic will be reduced by 6.6 dB if routed on top of the filter rather than referenced to the solid plane (as for the case $J28$–$J30$).

The time domain simulation is run by the Link Path Analyzer ver. 1.12 (tool developed at the Missouri University of Science and Technology, EMC Laboratory in Rolla, http://emclab.mst.edu/ in collaboration with the University of L'Aquila UAq EMC laboratory, http://orlandi.ing.univaq.it/uaq_laboratory/index.html). The results are provided in Figure 4.29 where the two output signals (through the pair $J28$–$J30$ and the pair $J40$–$J42$) are compared with the input. The peak voltage of the output common mode signal is reduced to 0.15 V for the $J28$–$J30$ case, and to 0.067 V for the $J40$–$J42$ case. The common mode reduction is obtained by (4.3):

$$20 \cdot \log_{10}\left(\frac{0.15}{0.067}\right) = 7.24\,\text{dB} \tag{4.3}$$

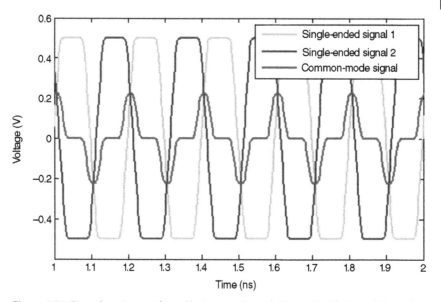

Figure 4.27 Time domain waveform. The two single-ended signals with 15 ps delay and the resulting common mode noise.

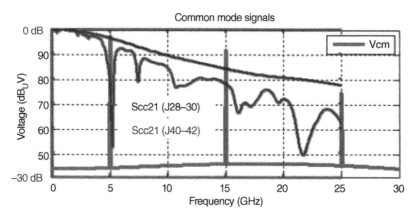

Figure 4.28 Spectrum of the common mode signal overlapped on the $|S_{cc21}|$ of the two measured differential pairs.

and it is very close to 6.6 dB associated with S_{cc21} at 5 GHz. Therefore, the quantification of the filter performances at the filter notch corresponds quite well to the reduction of the overall common mode time domain clock signal.

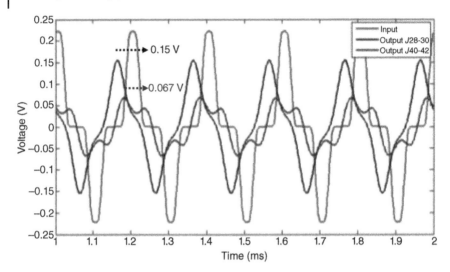

Figure 4.29 Time domain common mode signal at both the input and the output of the two differential trace pairs J_{28}–J_{30} and J_{40}–J_{42}.

4.3.3 Time Domain Common Mode Measurements

The performances of the designed common mode filter are validated also by time domain measurements. The experimental setup is shown in Figure 4.30. A signal generator (ANRITSU MP1763C) is used for generating the two opposite single-ended clocks (CLK1 and CLK1) at 5 GHz. A signal (CLK2) at the same frequency is employed as trigger signal for the receiving scope (Tektronix DSA8200 with the 80E40 high-frequency sampling module). Two equal phase shifters (PS1 and PS2) are connected at the output of the signal generator for minimizing, at first, any unbalances in the two paths, and then for generating a delay (and thus a common mode) between the two single-ended clocks CLK1 and CLK1.

The two single-ended clock signals are originally connected directly to the oscilloscope, without passing through the microstrips on the PCB. The signals obtained are shown in Figure 4.31. A 159 mV peak–peak common mode signal is observed. Then the first phase shifter 1 (PS1) is adjusted for minimizing the received CM noise, as from the results in Figure 4.32; the added extra length by PS1 allows achieving a minimum value of 77.1 mV. This residual noise can come from other sources of imbalance in the system rather than just a simple delay, whereas the phase shifters are effective only for delay kind of unbalance.

The second step is to include the DUT. Initially, the microstrip pair *J*28–*J*30 (no filter) is included in the setup as shown in Figure 4.30. The visualized peak–peak CM is 55.5 mV, as shown in Figure 4.33, lower than the original 77.1 mV

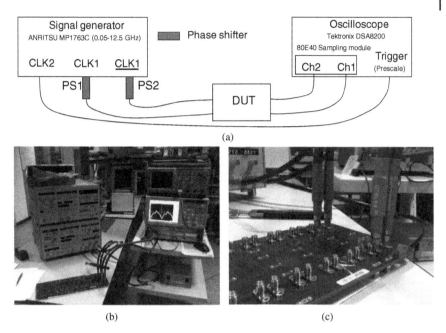

(a)

(b)   (c)

Figure 4.30 (a) Experimental setup for the time domain measurements. (b and c) Details of the setup.

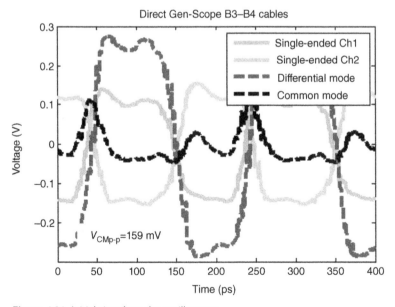

Figure 4.31 Initial signals at the oscilloscope.

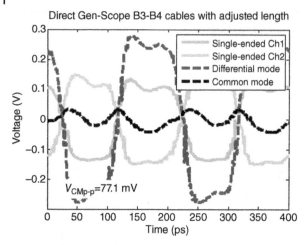

Figure 4.32 Minimized CM signal after adjusting one of the phase shifters.

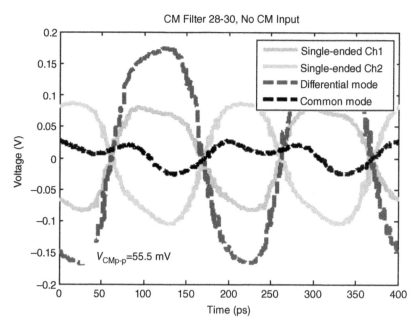

Figure 4.33 CM signal after the insertion of the DUT (J_{28}–J_{30} differential pair).

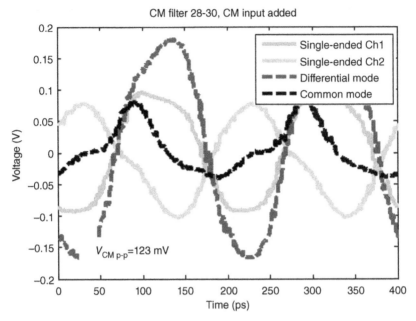

Figure 4.34 CM signal for the case $J_{28}–J_{30}$ increased by moving the phase shifter 2.

due to the loss of the PCB microstrip pair. The second phase shifter (PS2) is moved to increase the CM shown at the scope, as from the resulting signals in Figure 4.34. The obtained 123 mV peak–peak common mode is the value to be reduced, applying the several filter configurations designed on the PCB, as shown in Figure 4.24.

The S-parameters of three differential pairs are measured and the $|S_{cc21}|$ behaviors are shown in Figure 4.35 (the differential pairs $J28–J30$ and $J40–J42$ were already measured, as from the results in Figures 4.25a and 4.26a).

The deepest notch occurs for the configuration $J40–J42$, characterized by the smallest trace to patch edge distance ($d_1 = 3.4$ mm).

The results in Figure 4.35 validate the preliminary conclusions drawn by analyzing the results of three-dimensional simulations in Figures 4.20 and 4.21. The closer to the patch edge the trace crosses the gap between adjacent patches, the deeper the filter notch. This can be observed by also looking at the data in Figure 4.35, where the configuration with smaller d_1 leads to a deeper notch at 5 GHz. The distance d_1 associated with the four filter cases in Figure 4.35 are given in Table 4.2.

Figure 4.36 shows the time domain measurement results of the common mode received at the oscilloscope for the five measured DUTs. The bar on the right-hand side of the figure highlights the peak–peak common mode voltage

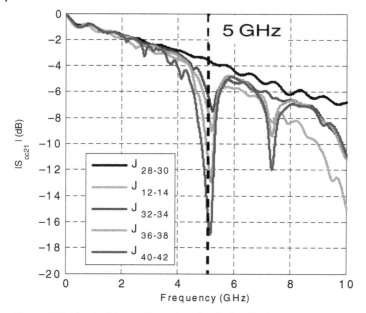

Figure 4.35 $|S_{cc21}|$ of the configurations J_{28}–J_{30} (no filter), J_{12}–J_{14}, J_{32}–J_{34}, J_{36}–J_{38}, and J_{40}–J_{42}.

Table 4.2 Numerical evaluation of the CM reduction.

Test site	d_1 (mm) (% of a)	$\|S_{cc21}\|$ (dB) at 5 GHz	CM reduction (dB)	Pk–Pk V_{CM} (mV)	Reduction from TD (dB)
J_{28}–J_{30}	/	−3.4	/	123	/
J_{32}–J_{34}	5.8 (41%)	−5.8	2.4	96	2.1
J_{12}–J_{14}	5.5 (39%)	−6.8	3.4	85	3.2
J_{36}–J_{38}	3.7 (26%)	−9.6	6.2	81	3.6
J_{40}–J_{42}	3.4 (24%)	−13.3	9.9	49	8

for each configuration, whereas their numerical values, together with the common mode reduction referred to the case without filter ($J28$–$J30$), are evaluated by (4.3) and then reported in Table 4.2. The CM reduction values are also shown in Figure 4.37. The data trend of the frequency domain measurements is similar to the trend from the time domain measurements; furthermore, the common mode reduction is inversely proportional to the trace to patch edge distance d_1.

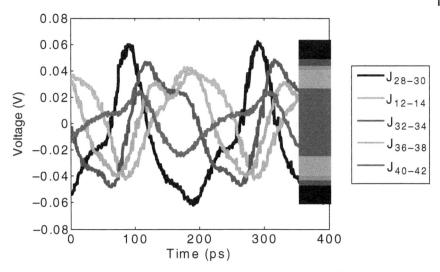

Figure 4.36 Measured common mode for the configurations $J_{28}-J_{30}$ (no filter), $J_{12}-J_{14}$, $J_{32}-J_{34}$, $J_{36}-J_{38}$, and $J_{40}-J_{42}$.

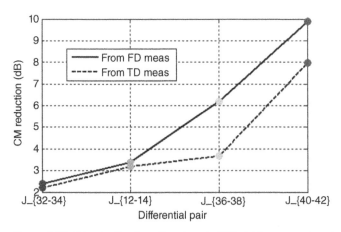

Figure 4.37 CM reduction from time domain (TD) and frequency domain (FD) measurements.

4.4 EBG-Based CM Filters: Design Approach

In this section, we will describe the evolution of CM filters based on resonant cavities to CM filters based on EBG structures. As a first step, we will introduce the reasons for the employment of CM filters based on EBG structures, for instance, the possibility to reduce the area occupied by the filter itself making it more appealing for the subsequent implementation on a real high-density PCB.

4.4.1 From Resonant Cavity CM Filters to EBG-Based CM Filters

As seen before, the current structure of a 5 GHz filter uses square patches ($a \times b$) whose typical dimensions are on the order of 10–15 mm (depending on the substrate permittivity). Taking advantage of the EBG properties already presented in Chapter 2, and employing the typical layout technique for reducing the noise propagation within power/ground plane pairs in PCB, the layout area required by CM filters based only on patch cavity resonators can be reduced. Each filter patch can be built as an EBG itself, allowing the filter size to be reduced significantly. This last property makes use of EBG structures much more attractive and realistic.

In the full-size CM filter structure employing only square patches, as seen in this chapter, the a and b dimensions are equal. The miniaturization process will change these two dimensions. Figure 4.38a shows the basic idea behind the reduction process: The first step consists in reducing the original a_1 dimension (associated with the configuration I) to a_2; the same first patch resonance (TM_{01} mode) is ensured according to (4.1) by the b_1 dimension that is unaltered. Thus,

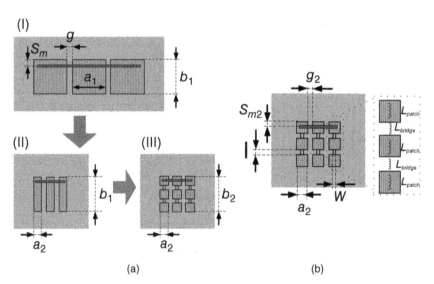

(a)   (b)

Figure 4.38 (a) Basic idea toward the miniaturization of the EBG-based common mode filters. (b) Miniaturized EBG-based filter configuration showing also the equivalent inductance model of one EBG cavity.

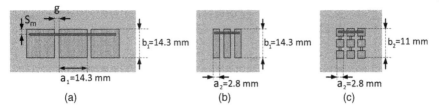

Figure 4.39 Comparison of the EBG resonant structure dimensions.

configuration II is obtained by rectangular patches. The second step is to reduce b_1 to b_2, always ensuring the same resonant frequency of the TM_{01} mode. This can be done following the properties of the EBG-type cavities, as described in Chapter 2. The small bridges placed between the smaller square patches of size a_2 increase the connection inductance between them, allowing lower resonant frequencies while maintaining a small patch size. The final miniaturized EBG cavity suitable for a CM filter is shown in Figure 4.38b.

Recently, additional studies were developed to validate the design strategy and to make this technique reliable to minimize the layout area required by this type of common mode filters. For this reason, in the remaining part of this section, the complete development process suitable for EBG-based CM filters will be explained in detail.

In order to compare the amount of board real estate required for the different options, a common mode filter designed for 5 GHz notch is taken as reference, as a consequence of the aforementioned parameters that have been customized for this target frequency [16]. Figure 4.39a shows the dimensions required for the common mode filter using the full-size design approach. Figure 4.39b shows the dimensions for the 3×1 solid patch approach, and finally, Figure 4.39c shows the dimensions for the 3×1 miniaturized EBG structure.

As one would expect, the results in terms of position are not identical for each filter configuration, but all of them are performing as a filter at the target design frequency of 5 GHz if we consider an attenuation threshold of -10 dB. Figure 4.40 shows the simulation results coming from a full-wave model for each approach. A line is drawn in Figure 4.40 at the level of -10 dB to allow the analysis of the filter notch bandwidth at this target attenuation. The 3×1 EBG filter gives the widest bandwidth, allowing tolerances in the frequency of the CM components to be filtered or in the dielectric properties of the substrate. In order to appreciate the gained beneficial effects, Table 4.3 shows the required board dimensions and the reduction from the original full-size design approach. The 3×1 solid patch configuration (Figure 4.39b) requires only about 24% of the board real estate compared to the original full-size EBG. The 3×1 miniaturized configuration (Figure 4.39c) reduces the board real estate even further to less than about 19%.

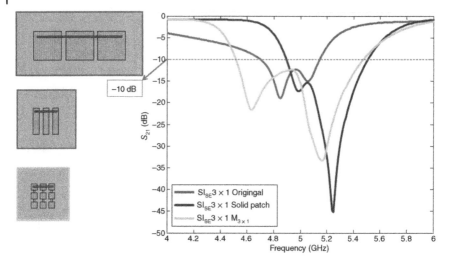

Figure 4.40 Comparison of the EBG notch filter performance depending on the miniaturization step. (S_{cc21} results taken from Ref. [24].)

Table 4.3 Comparison of PCB real estate requirements for different EBG configurations with a 5 GHz target notch frequency.

EBG configuration	Total surface (mm²)	Relative size required
Full size	650.65	–
3 × 1 solid patch	157.3	24.71%
3 × 1 miniaturized	121	18.6%

4.4.2 Synthesis Procedure of EBG-Based CM Filters

As easily understandable from the above discussion, the 3 × 1 miniaturized EBG configuration is the most desirable one, from both the size viewpoint and the bandwidth of the notch filter. This section will describe how to initially design the EBG filter. For this purpose, Figure 4.41 shows the general geometry that will be employed in the process under study; this can be done without loss of generality since these steps are applicable to other geometries also. It is important to notice that parameter B will be in all the cases smaller than the length used in the full-size EBG.

Figure 4.41 Miniature EBG filter dimensions.

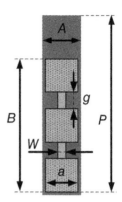

For the initial calculations, we will use parameter P as the relevant dimension to locate the frequency of the desired notch according to the following expressions:

$$f_{TM_z,01} = \frac{c_0}{2\sqrt{\varepsilon_r}}\sqrt{\left(\frac{1}{P}\right)^2} \tag{4.4a}$$

$$P = \frac{c_0}{2\sqrt{\varepsilon_r}f_{TM_z,01}} \tag{4.4b}$$

The inductance of N patches $(a \times a)$ and $N-1$ bridges $(g \times w)$ in Y-direction is written as follows:

$$L_{tot,Y} = \mu_0 d \frac{P}{a} = NL_{patch,Y} + (N-1)L_{bridge} \tag{4.5}$$

where d is the distance of the patches from the underneath reference plane. The patch inductance can be found from Equation (4.6) (taking into account the simplified case of square patches):

$$L_{patch,Y} = \mu_0 d \frac{b}{a} = \mu_0 d \frac{a}{a} = \mu_0 d \tag{4.6}$$

Because

$$L_{tot,Y} = \mu_0 d \frac{P}{a} \tag{4.7}$$

then

$$L_{bridge} = \frac{\mu_0 d(P/a) - N\mu_0 d}{(N-1)} \tag{4.8}$$

In (4.8), the number N of patches can be arbitrarily fixed. The sensitivity analysis in Ref. [25] has shown that when $N = 3$, the specific ratio $\lambda/a = 0.105$ can be taken into account, where λ is computed from Equation 4.9:

$$\lambda = \frac{c}{f_{TM_z,01}\sqrt{\varepsilon_r}} \tag{4.9}$$

At this point, the bridge inductance can be computed by considering the bridges similar to a microstrip structure. Equation 4.10 [1] can be used to calculate the bridge width (w) and length (l):

$$L_{bridge} = \begin{cases} l(60/c_0) \ln\left((8d/w) + (w/4d)\right), & w/d \leq 1 \\ l(120\pi/c_0)\left[(w/d) + 1.393 + 0.667 \ln\left((w/d) + 1.444\right)\right]^{-1}, & (w/d) > 1 \end{cases} \tag{4.10}$$

In fact, once the value for bridge inductance, calculated from Equation 4.8, has been found and after the dielectric thickness (d) has been given, the w and l values can be easily calculated.

It must be remarked that many combinations of w and l can exist that will give the same inductance. Either of the two parameters can be chosen, and the other computed to give the desired inductance. For this reason, the designer has to take into account the best trade-off between the values of these parameters; so in order to provide a useful tool for this aim, an optimized design workflow will be presented in the remaining part of this section.

4.4.3 Equivalent Circuit Analysis and Refinement of the Initially Synthesized EBG Filters

After having covered the main design workflow involved in the dimensioning of the EBG resonant cavity to be used as CM filters, some other aspects must be considered. It must be said, for the sake of completeness, that the calculations provided in the previous section by the closed-form equations are approximate, so obtained notch frequency should be then refined with a numerical tool. Let us start from the geometry in Figure 4.42 where an equivalent single-ended trace is used to simulate the propagation of the CM component. Given the

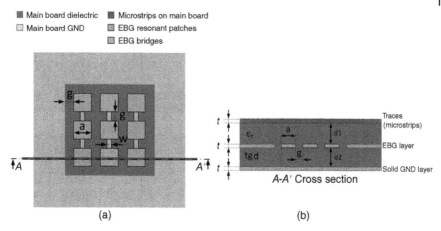

Figure 4.42 (a) Initial EBG filter design obtained with the design procedure (included in a larger PCB) reporting all the geometrical parameters of interest. (b) Cross-sectional view of EBG filter ($d_1 = d_2/2 = 0.2$ mm, $\varepsilon_r = 4.4$, $t = 0.017$ mm).

dielectric thicknesses equal to $d_1 = 0.2$ mm, $d_2 = 2 \cdot d_1$, the dielectric constant $\varepsilon_r = 4.4$, and setting the bridge sizes as $w = 0.4$ mm and $g = 1.3$ mm, Equations 4.4b–4.10 are applied obtaining $a = b = 2.8$ mm. When a three-dimensional electromagnetic full-wave analysis [18,26] is performed to check the notch frequency, it was found that the notch occurs at 4.47 GHz rather than the target frequency of 5.0 GHz. Therefore, further analysis should be performed to match the design specifications.

As an alternative to the computational burden of a three-dimensional full-wave simulation, a circuit analysis can be developed. The desirability of a circuit analysis model comes mainly from the fact that a proper tuning of the filter notch very close to the desired target frequency could involve the execution of many three-dimensional full-wave models, which can take a very long time to bring the designer toward the desired result.

The starting point for an effective circuit approach is the consideration of the path the common mode return current takes. Based on this, a model such as the one depicted in Figures 4.43 and 4.44 can be developed. Considering Figure 4.43, the CM return current flows

- on the EBG patches when the trace is over them,
- on the solid GND layer when the trace crosses the gap.

Because of this, pairs of ports (denoted by P_1 and P_2) can be set and placed as in Figure 4.43 in order to study each EBG section separately.

The analysis of each single EBG section was performed with both a full-wave tool [18,26] and a cavity resonance tool [27], based on the theory in Ref. [28].

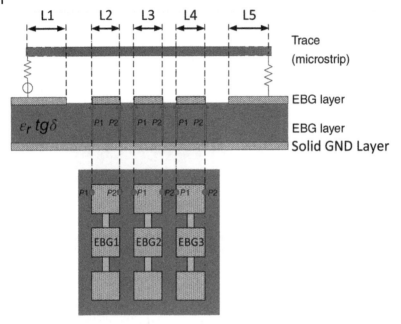

Figure 4.43 Return current path for the designed EBG filter of Figure 4.42.

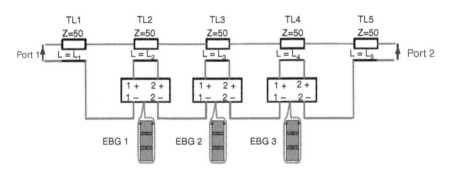

Figure 4.44 Equivalent circuit for the EBG-based CM filter under study.

The full-wave analysis was performed considering the actual geometry of the EBG patch and bridge, while the cavity resonance analysis was performed considering an equivalent rectangular patch of length $P = 14.3$ mm as in (4.4a) and width $a = 2.8$ mm. Both types of analysis allows one to obtain S-parameters data describing the EBG behavior. Since each section of the overall EBG filter is identical to the other, the S-parameter data will be identical for all the three sections. So, once we have the individual EBG S-parameters, the equivalent

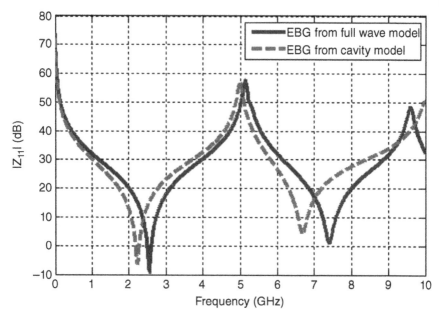

Figure 4.45 $Z_{11}|$ of the EBG cavity computed by a full-wave model and by the cavity model applied to the equivalent solid plane cavity. (Data from Ref. [24].)

circuit in Figure 4.44 can be built. This circuit includes *S*-parameter blocks for of each EBG sections and transmission lines for modeling the gaps. The circuit analysis can be performed with any of the existing circuit CAD tool, such as ADS, HSPICE, etc. Figure 4.45 shows the input impedance $|Z_{11}|$ at Port 1 of a single patch computed by the full-wave and by the cavity model tools.

Note that the frequency of the first resonant mode (TM_{10}) is very similar; the cavity model provides a precise value of 5 GHz, whereas the full-wave analysis provides a resonant frequency of 5.14 GHz. Since the EBG cavity can be efficiently simulated using the cavity model analysis, the *S*-parameter block computed by the cavity model is included in the circuit model in Figure 4.44. The equivalent circuit is evaluated and the results are shown in Figure 4.46. They are also compared with the results from the full-wave simulation of the entire structure in Figure 4.43.

The resonant frequency obtained by the calculation of the equivalent circuit is 4.63 GHz, and it is 4.47 GHz using the full-wave analysis. These values allow one to quantify the error of the approximation given by the equivalent circuit against the calculation of the EBG as an equivalent solid cavity. The error is found to be less than 5%. Furthermore, the agreement between the two curves in Figure 4.46 is quantified using the Feature Selective Validation tool [21–23,29,30]. The Grade and Spread associated with the ADM evaluation are computed for this case, and

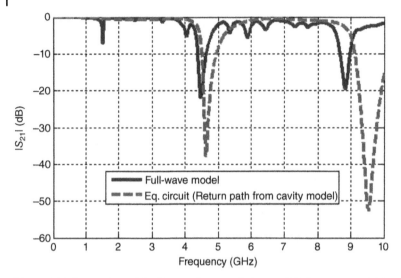

Figure 4.46 Comparison of the filter performance in terms of $|S_{21}|$, responses obtained from a full-wave simulation and from the equivalent circuit model. (Data from Ref. [24].)

the values are as follows: Grade = 6 and Spread = 4. A Grade = 6 is not normally acceptable; however, in this case, it is due to the poor agreement above 8 GHz, which is outside the frequency range of concern.

Studying the developed equivalent circuit model helps to understand that the impact of the trace crossing the gaps between adjacent EBGs is a shift of the EBG first resonant mode toward a lower frequency. This shift Δf can be quantified, so the overall design can be adjusted in order to move the notch frequency back to the desired location.

$$f_{actual} = f_0 - \Delta f \qquad (4.11)$$

where f_{actual} is the frequency of the notch whose value stems from the simulations and f_0 is the nominal design frequency. At this point, the filter can be redesigned with a new initial target notch frequency easily predictable using the following relationship:

$$f_0^{new} = \left(1 + \frac{\Delta f}{f_{actual}}\right) \cdot f_o \qquad (4.12)$$

The EBG design workflow can be now reviewed and updated as follows:

- f_{actual} is computed by (4.11);
- f_0^{new} is computed by (4.12);
- f_0^{new} is used in (4.4b)–(4.10).

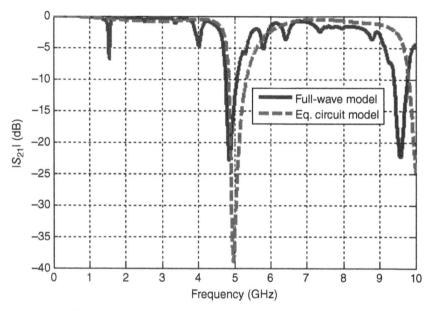

Figure 4.47 Filter performance in terms of $|S_{21}|$ of the refined geometry from a full-wave simulation and from the equivalent circuit model. (Data from Ref. [24].)

For the example in Figure 4.43, one obtains for a design frequency $f_0 = 5$ GHz, $f_{actual} = 4.63$ GHz, $f_0^{new} = 5.4$ GHz, and the new patch size $a = b = 2.55$ mm keeping the other geometry parameters w and g constant. Hence, the length of the equivalent solid plane cavity is $P = 13.24$ mm. The results from the full-wave simulation and the equivalent circuit model are compared in Figure 4.47. The notch frequencies are 4.87 and 4.97 GHz for the complete model simulated with the full-wave solver and the equivalent circuit model, respectively. Therefore, the proposed design workflow based on the formulas (4.4b)–(4.10) is able to predict the notch frequency with an error less than 5%. The $|S_{21}|$ behavior from the full-wave result includes some additional notches at 1.55 and 4.02 GHz, due to the resonant behavior of the surrounding solid plane area, which is inherently present as can be seen from Figure 4.43.

4.4.4 Optimum Geometrical Design

A solid rectangular patch of size $P \times a$ (as the dotted line in Figure 4.48) could be designed to resonate at a predefined frequency f_0, which corresponds to the design frequency to the common mode harmonic to be filtered. According to the previously mentioned workflow, the EBG filter design technique reduces the rectangular cavity area from $P \times a$ to $B \times a$, where $B < P$. Using a sequence of

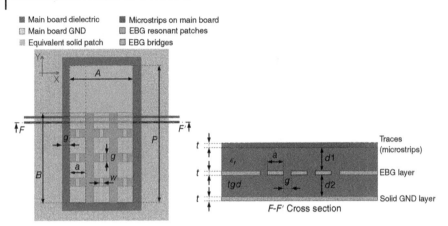

Figure 4.48 EBG filter dimensions and its equivalent pair plane structure (dashed contour) and the equivalent enlarged plane pair structure (dotted contour).

three adjacent vertical rows separated by the distance g leads to an overall layout area given by $B \times A$ with $A = 3a + 2g$.

According to what already done previously, the development of the EBG cavity follows the guidelines expressed in Section 2.2 and in Chapter 3 along with what stated also in this chapter (in the previous section). In particular, the calculation of the total inductance of the planar EBG structure along the Y-direction is done in agreement with (4.5).

Referring from now on to the example in Figure 4.8, P is once again the equivalent solid cavity length and it is calculated from (4.5), $L_{patch,Y}$ is the EBG patch inductance along the Y-direction, L_{bridge} is the bridge inductance, M is the number of patches along the Y-direction, and d is the dielectric thickness. Considering a parallel plate transmission line approximation, the $L_{patch,Y}$ value can be calculated according to (4.6), whereas the L_{bridge} can be computed assuming the bridge is a microstrip-like geometry, according to the formulation in (4.8).

A first design constraint is immediately derived from (4.8) in order to get a positive value for the variable L_{bridge}:

$$\left(\frac{P}{a} - M \right) > 0 \tag{4.13}$$

The bridge inductance calculated as in (4.8) depends on a, d, and ε_r. Once they are fixed (either by other constraints or by an arbitrary choice), the bridge geometry can be defined by solving (4.10) for either w or g. Assuming to evaluate g as final parameter, the EBG design has four degrees of freedom (a, w, d, ε_r) to calculate g.

Table 4.4 Simulation models to test first resonant notch.

Models	a (mm)	w (mm)	d (mm)	g (mm)
Model 1	3	1.20	0.05	2.54
Model 2	3	0.6	0.4	2.66
Model 3	6	4.2	0.4	1.37
Model 4	6	2.4	0.05	0.66

At the end, it is possible to design an infinite number of EBG for the desired first resonance frequency value. The key point of this section is to search for some constraints capable of reducing the range of parameters to be selected in order to get the most accurate design in terms of filtering frequency.

It is important to underline that following the general design procedure just explained, there are several sets of possible parameters available for the same target frequency—although depending on their selected specific values, the effectiveness of the design can reach different levels.

To check the effects of the wide range of parameter selection, four different 4×1 EBG models are designed for a desired filter frequency of 2.5 GHz. The models are synthesized setting $\varepsilon_r = 4.4$ and a, w, and d as listed in Table 4.4. The bridge length g is obtained by inverting the relationships in (4.10). These four models are simulated using the full-wave modeling environment CST Studio Suite. The computed $|Z_{11}|$ parameters are compared, as shown in Figure 4.49. The excitation port is located at one of the end patches of the EBG 4×1 array.

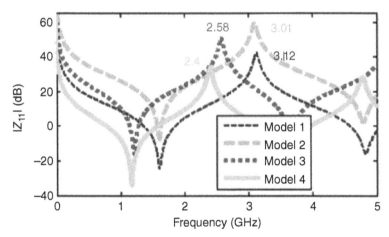

Figure 4.49 $|Z_{11}|$ of the four test EBG cavities designed to resonate at 2.5 GHz, and computed by a full-wave model. (Data from Ref. [25].)

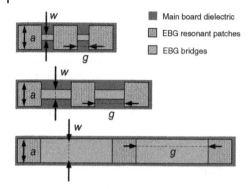

Figure 4.50 Three EBG designs based on the same patch size, varying the bridge width and length.

As it can be seen from the results in Figure 4.49, the first resonant frequency values are different for each model. Model 3 provides the resonant frequency closest to the 2.5 GHz design value. These differences are due to the approximations introduced by the analytical procedure; thus, not all the possible designs lead to the same design accuracy. In order to define some limitations in the use of the developed EBG design synthesis, (4.8) and (4.13) are analyzed and some preliminary considerations can be summarized as follows:

1) A positive bridge inductance value ($L_{bridge} > 0$) requires that (4.13) is satisfied. Thus, the maximum value of a can be calculated considering (4.13).
2) The bridge width w has an associated maximum value such that $w \leq a$. The limit value for which $w = a$ corresponds to a solid plane, as shown in Figure 4.50.

After these considerations, several geometries are investigated to achieve the desired notch frequency of 2.5 GHz. The synthesis procedure is applied once again, as done for the cases in Table 4.4, but varying a from 3, 4.3, 4.776 to 6 mm, d from 0.05, 0.15 to 0.5 mm, and w/a from 0.125, 0.2, 0.4, 0.7 to 1, in order to check the effects of the EBG parameters on the EBG first resonant frequency. The last value for $a = 6$ mm is calculated from (4.13), while the other a values have been chosen randomly. The bridge length g is calculated by inverting the relationships (4.10). The parameter groups employed are summarized in Table 4.5.

From the enumerated parameters, 60 different 4×1 EBGs are designed and simulated with a full-wave solver. The obtained first resonant frequency values are extracted and plotted as in Figure 4.51 as function of a/λ, where λ is defined according to (4.9) as in (4.14). This plotting format helps to identify the relationship between the patch size and the design frequency:

$$\lambda = \frac{c_0}{f_{10}\sqrt{\varepsilon_r}} \tag{4.14}$$

Table 4.5 Simulation models to test first resonant notch for 4×1 EBGs.

d (mm)	Frequency (GHz)	a (mm)	w/a	g (mm)
	2.5	3	0.125	Calculated
0.05		4.3	0.2	
0.15		4.776	0.4	
0.4		6	0.7	
			1	

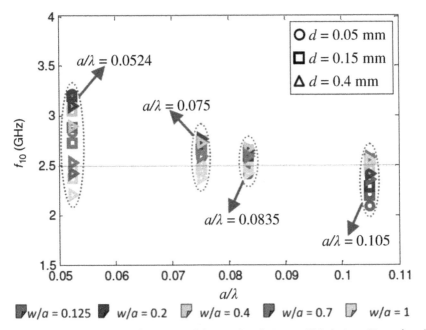

Figure 4.51 First resonant frequency of the simulated 60 4×1 EBG designs. (Reproduced from Ref. [25] with permission of IEEE.)

In order to identify the better a/λ ratio value, the error percentage between the design filter frequency $f_0 = 2.5$ GHz and the EBG first resonance from simulation (f_{sim}) is evaluated following the formula in (4.15). The obtained error is plotted in Figure 4.52 for each model.

$$Error(\%) = \frac{|f_0 - f_{sim}|}{f_{sim}} \times 100 \qquad (4.15)$$

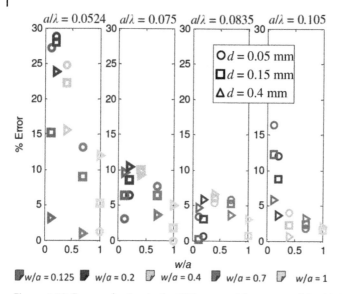

Figure 4.52 Error evaluation in the EBG resonant frequency of the 60 4×1 EBG models. (Reproduced from Ref. [25] with permission of IEEE.)

As can be seen from Figure 4.52, the better results are obtained for $a/\lambda = 0.0835$ (Error $< 10\%$), and it can be thought acceptable, while the worst case is $a/\lambda = 0.0524$ with an error around 30%.

Choosing $a/\lambda = 0.0835$ as a reference value, additional constraints on the relationships between a/λ and the patch number M can be studied. Note the fact that (4.8) can be rewritten as in (4.16), thus obtaining that the value $\mu_0 d$ should be constant for a constant dielectric thickness d:

$$\mu_0 d = \frac{(N-1)L_{bridge}}{((P/a) - M)} = const. \tag{4.16}$$

Thus, (4.17) can be defined setting different patch number M:

$$\mu_0 d = \frac{(M-1)L_{bridge-M}}{((P/a_M) - M)} = \frac{(N-1)L_{bridge-N}}{((P/a_N) - N)} \tag{4.17}$$

where M and N are the different patch numbers (i.e., $N = 3$, $M = 4$). It can be reasonably assumed that the contribution of bridge inductance to (4.5) is proportional to the patch and bridge number, and then the following condition is set:

$$L_{bridge-N} = \frac{M}{N} L_{bridge-M} \tag{4.18}$$

Table 4.6 Parameters of the simulation models run to test the first resonant notch.

d (mm)	Frequency (GHz)	*a* (mm)	*w/a*	*g* (mm)
0.05	2.5		0.125	Calculated
0.15		8	0.2	
0.4		6	0.4	
		3	0.7	
			1	
	5		0.125	Calculated
		4	0.2	
		3	0.4	
		1.5	0.7	
			1	
	8		0.125	Calculated
		2.5	0.2	
		1.87	0.4	
		0.93	0.7	
			1	

Thus, as an example, $L_{bridge\text{-}3}$ can be obtained from (4.18) as shown in (4.19):

$$L_{bridge-3} = \frac{4}{3} L_{bridge-4} \tag{4.19}$$

At this point, because of this useful relationship, the new value for the patch size a with $N = 3$ (a_3) can be calculated by using the formulation hereafter reported:

$$a_N = \frac{a_M NP(M-1)}{M(NP - P + a_M N^2) - a_M (M^2 N - M^2 + N^2)} \tag{4.20}$$

The value for a_3/λ, thus for $N = 3$, is 0.105. The validity of this last relationship has been in-depth investigated in Ref. [25]; in particular, a wide set of additional full-wave models have been simulated in order to show the effectiveness of the fundamental patch size constraint set by (4.20). Just to briefly report the main results obtained, it can be said that 135 simulations have been arranged according to the parameters listed in Table 4.6.

The a values obtained from the application of (4.20) are 6, 3, and 1.87 mm for the three filter frequencies 2.5, 5, and 8 GHz, respectively. These frequency values are directly associated with the specifications of typical communication rates in the Gbps digital interconnects. Without loss of generality, the other patch size values are chosen randomly.

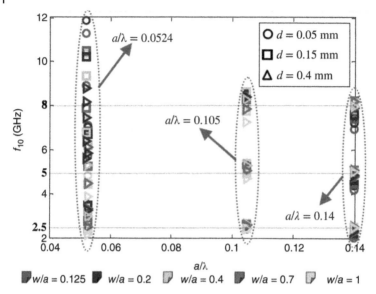

Figure 4.53 Resonant frequency of the 3 × 1 EBG models designed for 2.5, 5, and 8 GHz varying a/λ. (Reproduced from Ref. [25] with permission of IEEE.)

The first resonant frequency values are plotted in Figure 4.53 as a function of a_3/λ, for the three considered frequencies: 2.5, 5, and 8 GHz. It is worth noting that the results related to the smallest $a_3/\lambda = 0.0524$ are spread along the frequency axes; thus, it is not possible to discern which frequency was intended. The value $a_3/\lambda = 0.105$ provides the best results and the minimum error regardless of the design frequency, always below 10%, as assumed from the derivation in (4.16)–(4.20). The work developed in Ref. [25] reports also the complete analysis of the errors for the three frequencies of interest by using error plots. In particular, the error is evaluated as in (4.15) and it is plotted for different w/a when the electrical length of a, a/λ, changes. From the analysis of such information, the most important item to note is that a best value for w/a cannot be identified starting only from the results obtained for the errors.

On the other hand, the limited error, which was computed for the case $w/a = 0.125$ and $w/a = 0.2$, is always below 7% for both $M = 4$ (as also shown in Figure 4.52 for $a/\lambda = 0.0835$) and $M = 3$ (for $a/\lambda = 0.105$), and can be considered as the optimum value. This conclusion does not diminish the other key points. As mentioned earlier in this section, the use of the EBG instead of a solid cavity allows a reduction of the filter layout area. Therefore, using a narrower bridge leads to a reduced g, starting from the computed L_{bridge}, reducing in this way the overall EBG layout area.

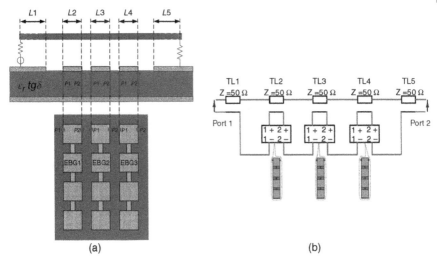

(a)   (b)

Figure 4.54 (a) Identification of the return current path for the EBG filter. (b) Equivalent circuit model using ADS.

4.4.4.1 Optimum Design Example

In order to show the effectiveness of the optimized design approach, it is also possible to apply the equivalent circuit previously introduced. Among all the models listed before, we chose a model coming from Table 4.5 accounting for a patch size equal to 4.776 mm and for a ratio $a/\lambda = 0.0835$. The bridge width is computed as $w = 0.955$ mm from $w/a = 0.2$. As an example, the dielectric thickness is fixed to $d = 0.15$ mm. The bridge length is obtained from the fundamental relationships in (4.10) as $g = 0.908$ mm. It is assumed that the dielectric layer is made out of FR4/epoxy material, which has a nominal dielectric constant of about 4.4, and that the metal thickness is 0.017 mm. The optimally designed filter structure has a geometry like the one shown in Figure 4.48 and an equivalent circuit model, whose topology in ADS [31] format is shown in Figure 4.54b.

The $|Z_{11}|$ results of this EBG structure from full-wave simulation are given in Figure 4.55. The approximated results are also obtained simulating the EBG as its solid plane pair counterpart of length $P = 28.6$ mm and width $a = 4.776$ mm, as done in Ref. [24] using a cavity model approach [32]. This result is also included in Figure 4.55. Both models predict accurately the resonant frequency at 2.5 GHz. The significant result here is that the use of a proper EBG cavity instead of the solid plane cavity reduces the required layout area from 136 mm^2 (P^*a) to 104 mm^2 ($4a^2 + 3ga$), in agreement with the target of a realistic practical design.

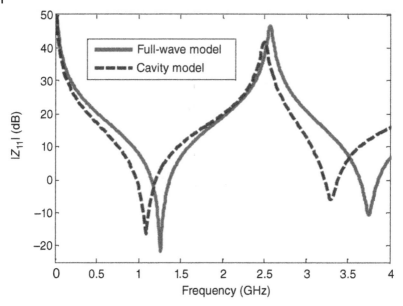

Figure 4.55 $|Z_{11}|$ comparison of the EBG cavity computed by a full-wave model and its solid plane counterpart computed by a cavity model. (Data from Ref. [25].)

The circuit model results are shown in Figure 4.56a, with the return path modeled using the two methods previously described. In this case, the predicted shift of the resonant frequency at lower frequency is negligible, so the designed EBG filter can be considered completed.

The results for the full-wave model of the complete geometry are also compared in Figure 4.56a. The notch frequency is at 2.43 GHz, compared to 2.46 and 2.52 GHz from the equivalent circuits with return path from the cavity model and from the full-wave simulation, respectively, as more clearly shown in Figure 4.56b. The error in the notch frequency is still below 5%, as expected from the error quantification provided in the previous section. The full-wave simulation results provide two additional features in the frequency spectrum of $|S_{21}|$, a notch at around 1.25 GHz, and the small notch at 2.25 GHz; they are due to the resonance of the large external ring surrounding the EBG filter section (see Figure 4.48). Even though the filter band is small, a larger bandwidth design can be done designing the EBG cavities with different resonant frequencies as first introduced in Ref. [24] (i.e., applying the proposed procedure for synthesizing three different EBGs). This additional design step can overcome the error given by the present analytical design and possible manufacturing uncertainties on geometrical (i.e., trace/bridge geometry) and electrical parameters (i.e., dielectric permittivity). This approach will be discussed in more detail in later sections.

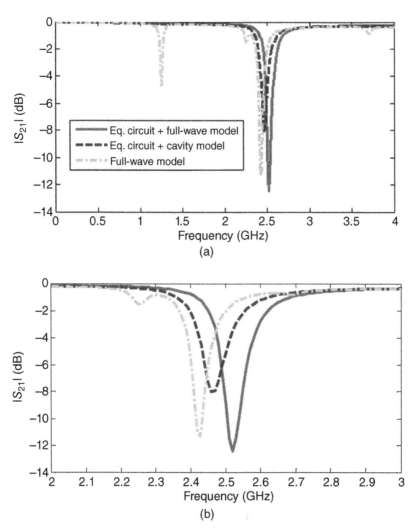

Figure 4.56 Filter performance in terms of $|S_{21}|$. (a) Comparison between the equivalent circuit model with the return path computed by a full-wave model and by the cavity model. (b) Zoom between 2 and 3 GHz. (Data from Ref. [25].)

4.5 Onboard EBG-Based Common Mode Filters: Typical Structures

As introduced in the preceding section and according to the actual state of the art in this field, the planar common mode filters based on EBG patch resonances have been demonstrated to be effective for the reduction of common mode propagation [2–4,11]. As already shown, in their original configuration, the EBG-based CM filters consist of microstrip pairs coupled with the EBG cavity. Sometimes, the implementation of such a kind of filter is done instead of accounting for the embedding of planar EBG structures deep inside the same stack-up of the main PCB; in this case, stripline traces need to be used. However, in all these cases, the inherent nature of these particular technological solutions gives rise to the so-called onboard EBG-based CM filters, since the filtering part itself is effectively built-in inside the main board, exploiting either the external layers or some internal layers. Other possible technological solutions are allowed when the implementation of the EBG filters is made apart from the main PCB, but in this case we will have, in contrast to the former solution, the so-called removable filters (which will be analyzed accurately in Chapter 6). However, for the purposes of this chapter, the focus is only on the "onboard" EBG-based CM filters. All these aspects concerning with their basic structure and the related design rules will be discussed.

4.5.1 Onboard EBG Structures for Common Mode Filters Development

In order to start listing the main aspects for which the development of an onboard EBG filter can be attractive, some initial considerations will be pointed out. First of all, one keypoint of the onboard EBG CM filters is the capability to include the filtering structures as a part of the main PCB manufacturing process. This provides a reduction of the costs with respect to other solutions implementing the CM filters as separate stand-alone components. Another keypoint is the simplicity of the design workflow that can be followed to properly size this devices, which can take advantage also from the achievement of quite good filtering performances. Moreover, since the filter structures are included inside the main PCB itself, there are no additional design considerations that an external component may require, for example, the transitions of the signal traces between the layers and the potential effects of vias.

As stated previously, any onboard EBG-based CM filter can be designed to either lie on the external layers of a PCB or be buried in the internal layers, depending on the needs of the designers during the development process. Even though the operating principle underlying the CM filtering properties would remain the same described in the previous sections, the design workflow must be adjusted to fit the currently available filter structures. There are at least two types of onboard EBG-based CM filters: those having a structure like the ones analyzed previously, where the EBG filter is laid out on the external PCB layers and coupled with microstrip differential traces; the other type, which is buried into the internal PCB layers and coupled with striplines instead, constitutes an alternative structure that could be also named "embedded" or "stripline" EBG CM filter.

(a)   (b)

Figure 4.57 Example test board accounting for a regular planar onboard EBG-based CM filter. (a) Perspective view of a 3D simulation model. (b) Photograph of a real manufactured test board.

4.5.1.1 Onboard EBG-Based CM Filters: Microstrip Structures

Looking from a macroscopic point of view, the general aspect of an example test board including a regular planar EBG-based CM filter and coupled with a microstrip differential pair is shown in Figure 4.57. Note the presence of four SMP connectors used for signals injection and the outlined shapes of the EBG resonant patches employed by the CM filter, which in this case are placed on the metal layer just below the microstrips. The photograph of Figure 4.57b shows an overall view of a real manufactured test PCB including such filters.

The general configuration together with some structural details of an example microstrip onboard CM filter can be seen looking at the diagram shown in Figure 4.58a. As easily seen, it is very similar to the previously analyzed configurations, since this example is also based on three EBG sections with three patches connected by the bridges. This choice is motivated by the fact that several significant properties have already been defined. The relevant geometry parameters are once again represented by the patch width (a), the gap between the patches (g), and the width of the interconnections between them (w), namely, "bridges." As said earlier, the filtering effect is achieved due to the resonant behavior of the cavity made by each single EBG patch and the solid plane reference below. For this reason, no vias should be placed within the patch area and no connection should be made between the patches and the reference plane in order to not prevent the correct resonating phenomena introduced by the filter itself.

Clearly, this is the simplest onboard EBG-based CM filter one can practically design, since it is laid out on the PCB outermost stack-up layers (top and bottom) but it offers quite good features without introducing great expenses or difficult layout efforts.

4.5.1.2 Onboard EBG-Based CM Filters: Stripline Structures

Normally, the basic planar EBG geometry is realized by a suitably designed patterned plane (see Figure 4.58) on top of a continuous one, creating a cavity

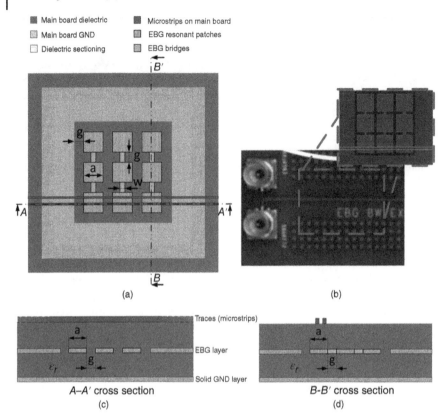

Figure 4.58 Geometrical details and stack-up structure of the EBG resonating cavity employed in the example onboard CM filter introduced in Figure 4.61. (a) EBG geometry. (b) Real main board top view identifying the placement of the EBG-based CM filter. (c and d) Stack-up structure highlighted from side view cross sections.

that behaves as illustrated in Figure 4.14 and gaining the already widely discussed properties. This cavity in its most general behavior shows a bandgap on the frequency spectrum of S_{21} ranging from a certain value f_{low} to another value f_{high}, according to what already stated by the guidelines expressed in Chapter 2 about the EBG resonating cavities. Despite this general behavior, the essential feature for the development of an EBG CM filter is the resonant frequency of the first notch occurring in the S_{cc21} response, since it defines the working frequency of the filter and then, in turn, it constitutes also the starting point for the design of the EBG filter geometry. However, if a third continuous layer is added on top of the EBG layer (as in Figure 4.59b), a new subcavity is obtained that behaves as accurately illustrated in Section 2.6. This last configuration has been introduced in Refs [33,34] as an "embedded EBG." It employs

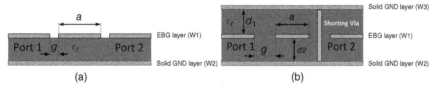

Figure 4.59 Stack-up of (a) planar EBG on outer layer and (b) embedded planar EBG placed on inner layers including also the shorting vias used to recover the bandgap (refer to what stated in Chapter 2).

shorting vias between the solid layers above and below the patterned one in order to obtain the same bandgap as if the patterned plane were on the outer layer, as discussed in Ref. [35].

Starting from the stack-up characterizing an embedded EBG resonant cavity, shown in Figure 4.59b, an EBG-based CM filter customized to include all the cases where the filter itself is buried inside the PCB stack-up can be constructed. This kind of EBG-based CM filter, exploiting in some way nested resonating patches, can be intended as the embedded version of the regular planar EBG and hence the solution to be used when stripline differential pairs are used in the PCB design. It can be built starting from the structure shown in Figure 4.59b and applying some structural modifications. The most important one is that, since we are interested only in the first resonance notch, there is no longer interest in the use of the shorting via. The resulting general embedded CM filter structure is shown in Figure 4.60. As can be seen from the figure, it is created by

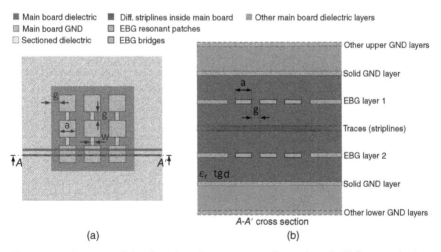

Figure 4.60 Geometrical details and stack-up structure of an onboard CM filter employing embedded EBG structures. (a) Top view of the EBG geometry from a cross-sectional cut plane located on the traces layer. (b) Side view of the stack-up structure showing all the significant filter layers.

using two patterned layers, one above and one below the traces and externally to these the usual solid GND reference layers. Because of its inherent structure, this solution will also be referred to in the rest of the chapter as "stripline" EBG-based CM filter. The features offered by such a kind of onboard filter solution are similar to the microstrip case.

From the design workflow viewpoint, a procedure quite similar to the one employed for the microstrip EBG-based CM filters can be applied also to the stripline cases. The stripline structure itself consists inherently of two reference planes, while the microstrip structure had only one. In order to reach a sufficient filtering action, it is required that both the references must be of EBG type, otherwise the major part of the return current will flow on the lower impedance path of the solid reference. In the more general situation, the EBG designs for each reference can be different, because the dielectric cavities in the test vehicle can have different thicknesses above and below the stripline layer. Starting from the same (equal for the two EBG structures) target frequency, the application of the previous synthesis procedure, exploiting the optimum parameter relationships presented in Ref. [25] and following the same steps as for the microstrip filter design case, leads to obtain an initial set of preliminary EBG dimensions.

At this point, a shift with respect to the originally set target frequency is experienced. This frequency shift is due to the application of the design rules (4.4a)–(4.10), which were formally deduced for the microstrip case only and for which air was supposed to be above the EBG, In this configuration of a stripline structure, the EBG cavity is effectively embedded within a homogeneous dielectric.

On the other hand, the parameters are found using the microstrip EBG design approach, since they can constitute a good preliminary approximation of the EBG geometry for stripline filters. However, this initial set of parameters need to be accurately refined. For this reason, the best practice is to make use of full-wave 3D solvers, as suggested in Ref. [36], in order to properly refine the initial approximations. Usually, the starting set parameters are not so far from the final ones (less than 12%), so they can be quickly tuned with a limited number of iterations.

4.5.2 Onboard EBG-Based CM Filters: PCB Implementation Aspects

When dealing with practical applications, it is not so unusual to get involved with mixed combinations of the two types of onboard filters. In fact, in some cases, due to strict routing constraints or due to the inherent application requirements, it is possible that for certain regions of the PCB under development, the most convenient choice would be to use regular EBG-based CM filters; on the other hand, there could also be other regions for which the employment of embedded filters is unavoidable in order to maintain the top/bottom layers free from EBG filters. Therefore, the overall PCB design may

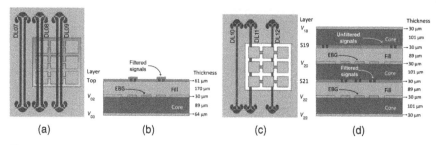

(a)   (b)   (c)   (d)

Figure 4.61 Example designs for microstrip and striplines EBG-based CM filters implemented on a real PCB for cross talk investigation purposes. *Microstrip solutions:* (a) Top view. (b) Cross section. *Stripline solutions:* (c) Top view. (d) Cross section.

require the employment of both the kinds of onboard filters. Naturally, it is responsibility of the designer to choose the best trade-off, taking into account different aspects and sometimes also considering experience-based figures of merits.

In a real-world implementation of such filters, where the system complexity leads to very high routing density, a limitation in the use of onboard CM filters is constituted by the fact that several trace pairs may cross the same filter with a very short separation distance. Therefore, we need to study the effects of the cross talk between the different pairs of traces.

A study to investigate in detail such cross talk issues was done in Ref. [37]. In that work, the regular EBG-based filter, which is the most straightforward implementation, was applied to differential microstrips, as shown in Figure 4.61a and b. The depicted figures show the actual layout of a real manufactured board that was employed to experimentally investigate the cross talk among adjacent microstrip differential pairs routed on the same EBG filter. This cross talk analysis considered both the microstrip EBG and the embedded EBG. In this latter case, the filters were intentionally allocated deeper in the stack-up, as depicted in Figure 4.61c and d. The generic stripline filter consists of two patterned layers above and below the differential traces; it must be pointed out that in this particular case, the return current flows on both the V_{20} and V_{22} planes, as highlighted in Figure 4.61d.

In order to show the effectiveness of both types of filters (with microstrip and with stripline) in a practical application but also the reliability of the design approach that is proposed in this chapter, the agreement between the predicted modeling results (obtained from proper 3D simulation models) versus the measured ones is considered next. In particular, Figure 4.62a shows the comparison of the common mode insertion losses for the case of a regular EBG-based CM filter accounting for the microstrip traces indicated as DL08 in Figure 4.61a. On the other hand, the same comparison is shown in Figure 4.62b for the case of an embedded stripline filter accounting for the differential pair

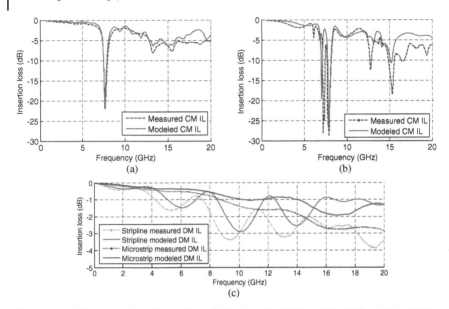

Figure 4.62 CM insertion loss for the filtered (a) microstrip DL08 and (b) striplines DL11. DM IL for both cases (c).

DL11 shown in Figure 4.61c. Note that for this specific application, the target frequency of the CM filters was set to 7.6 GHz and so the design workflow was customized for that target frequency. As mentioned earlier, the synthesis procedure is essentially the same for both the microstrip and the stripline filters. As seen in the obtained results, the main filter notches are effectively placed around the desired frequency for both the microstrip and the stripline EBG filters.

While cross talk can be complex, some observations can be made initially. First, the traces can pass over the patches following three configurations: left, middle, and right. Then, each cross talk scenario can be denoted by the pair of letters (L = left, M = middle, and R = right) indicating the position over the EBG of the aggressor and of the victim lines. As seen in Figure 4.63a and b, note that the CM near-end (NEXT) and far-end (FEXT) cross talk are generally lower for the stripline filters having left-to-middle (L-to-M) coupling, but the opposite behavior holds for the differential cross talk. On the other hand, for the microstrip cases, there is a very low variation between the cross talk scenarios L-to-M and L-to-R. Clearly, from a PCB design point of view, the CM FEXT is the most important quantity to be evaluated, since it can show if the CM filtering action on one differential pair can negatively affect other pairs crossing the same EBG.

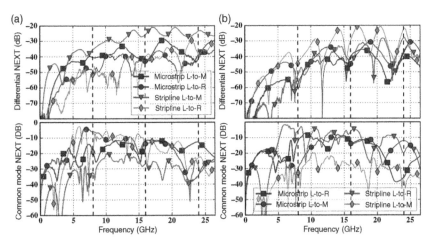

Figure 4.63 (a) Near-end coupling between filtered pairs over different locations on the EBG patches: left (L), middle (M), and right (R). (b) Far-end coupling between filtered pairs over different locations on the EBG patches: left (L), middle (M), and right (R). (Data from Ref. [37].)

Furthermore, from Figure 4.63b, note the near 0 dB CM coupling for the stripline L-to-R cross talk configuration at the EBG resonant frequency provides results indicating that multiple differential pair crossing of the same EBG is not optimum.

4.6 Additional Design Considerations

In this section, the focus will be put on some interesting points to be considered for the development of effective onboard EBG-based CM filters. One of the topics covered in this section will be the study of further spatial miniaturization techniques useful to construct filters that can be more appealing for integration on high-density PCBs. The presented miniaturization techniques will involve not only the layout geometry redesign but also specific stack-up solutions, as done in Ref. [38].

4.6.1 Further Miniaturization Techniques

Until now, the illustrated design workflow involved only the sizing of the resonant patch cavities and exploiting of the EBG-shaped reference planes in order to build effective CM filters. In Section 4.4, the EBG-based CM filter's dimensions were miniaturized in order to be more attractive from a layout viewpoint. However, the ever-increasing demand for higher density boards make the centimeter scale still not appealing.

For this reason, further miniaturization techniques are required to reach sensible levels of used surface reduction enabling EBG-based CM filters to be of interest for practical PCB design applications and at the same time for easy application.

In this section, we will present some surface reduction techniques based on different operating principles that can allow several degrees of freedom for the reduction of the surface occupied by EBG-based CM filters. In the first case, the reduction of occupied space is made through the elimination of less important elements included in the resonant EBG geometry. In the second case, the approach will consist in the optimization of the bridge geometry in order to get the same values of inductance calculated through the basic general design approach but using less space. In the third proposed scenario, there will be a redesign of the regular EBG-based filter structure in such a manner as to have the same filtering performances but changing the layout of the EBG resonant cavity, in particular exploiting a proper redesign of the stack-up structure.

Other important points involved in the development of such miniaturization techniques will be linked to the possibility of strategies to reduce the powerplane coupling effects induced by the presence of the miniaturized EBG filters. This is very important because it is expected that introducing such devices will not bring any kind of perturbation inside the original PCB design in terms of EMI issues. Toward this aim, a strategy will be presented that employs a certain number of vias used to stitch the surrounding of the filter structure. In this manner, it is expected that the field coupling outside the EBG-based CM filter area will be drastically reduced enhancing the appealing of such devices.

As mentioned previously, another aspect characterizing the practical employment of EBG-based CM filters is the cross talk performance. Because of this, also for the miniaturized EBG filters a brief cross talk analysis will be outlined in order to see if there is any beneficial effect coming from the employment of the miniaturized EBG geometries. Also, the performance of the miniaturized filters from the viewpoint of radiated power will be briefly addressed since this last aspect is of paramount value when dealing with a real PCB design. In fact, the additional presence of the filter should not introduce further radiation sources. It is expected that the total radiated power should be reduced using a proper filter structure. So the three solution will be compared from this point of view.

For an initial design, the following example geometry is illustrated in Figure 4.58. According to the design procedure described in previous parts of this chapter, the starting EBG-based filter has been designed as shown in Figure 4.64. It can be considered the "original" configuration. It is characterized by the indicated geometrical dimensions since it has been designed for a target frequency of 8 GHz and based on the reported stack-up accounting for a dielectric substrate with $\varepsilon_r = 3.6$ and $tg\delta = 0.012$.

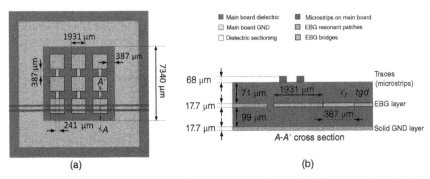

(a)   (b)

Figure 4.64 Reference EBG-based CM filter design, namely, "original" configuration. (a) Top view including the geometry dimensions. (b) Stack-up data.

4.6.1.1 Reduction of the Number of Patches

The first idea to reduce the required EBG space on the PCB is the elimination of the central patch in the three-patch "original" configuration since it has been found it is not essential for the construction of a resonant cavity. The central patch is then replaced by a longer bridge linking the remaining two patches. Since this bridge is narrower than the initial patch, a larger inductance will be associated with the newborn two-patches cavity. This will shift the notch of the EBG filter to lower frequencies with respect to the "original" EBG case.

However, the EBG filter designer can restore the desired notch frequency by reducing the length (and hence the inductance) of the bridge. This is the basic principle underlying this first surface reduction technique. After a preliminary tuning stage of the EBG geometry by adjusting the bridge length, the new EBG configuration, shown in Figure 4.65a and indicated as "reduced number of patches," is obtained. The comparison of the S_{cc21} transfer function for both this last filter configuration and the one where only the central patch was eliminated, namely, "no central patch," is shown in Figure 4.65b. In this figure, the two curves have also been compared with respect to the "original" case baseline.

Note that in the new EBG configuration, the distance of the two remaining patches has been reduced by a distance equal to 3.8g, g being the variable indicating the gap between two patches in the classical EBG filter design approach. Globally, the filter size has been reduced from 7.34 mm (of the 3 × 3 solution) to 5.87 mm, with a saving of PCB surface of approximately 20%. The notch frequencies in the "original" and "reduced number of patches" cases match very well and are located at the desired target frequency.

4.6.1.2 Modification of the Bridge Geometry: "Meandered" Configuration

The second configuration employed for further miniaturization purposes is the so-called meandered configuration, first presented in Ref. [38]. With this configuration, the goal is to further reduce the allocated PCB surface using

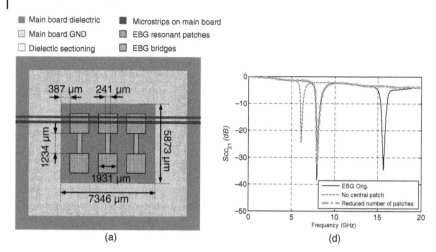

(a)   (d)

Figure 4.65 EBG-based CM filter "no central patch" configuration. (a) Top view. (b) $|S_{cc21}|$ response.

a folding of the bridges between the patches, as shown in Figure 4.66a. From a geometrical point of view, it is clear that there is an inherent relationship between the folding of the bridge and the reduction of its width in order to avoid undesirable electric short circuit of the bridge with itself or the patches. Because of this, the width of the bridges has been settled to the minimum value that can be obtained with a predefined PCB technology: in this case 127 μm.

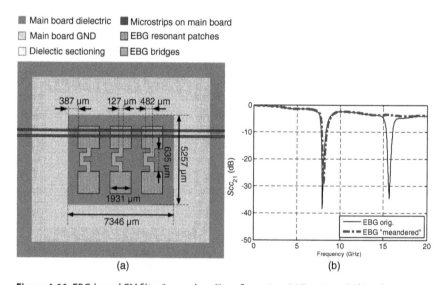

(a)   (b)

Figure 4.66 EBG-based CM filter "meandered" configuration. (a) Top view. (b) $|S_{cc21}|$ response.

Subsequently, a tuning procedure can be applied to the "meandered" EBG in order to have the S_{cc21} notch at the same frequency as for the "original" reference configuration. In this specific case, only one bridge folding was used to reduce or increase the total length of the bridge between the two patches by adjusting the length of the two horizontal parallel meander branches. Figure 4.66a shows the details of this final configuration where the meanders horizontal segments are 482 μm long while the vertical segments are 254 μm long. The $|S_{cc21}|$ transfer function is shown in Figure 4.66b, where a good match between the notches of the two configurations at the design frequency of 8 GHz is clearly visible. The reduction of space between the patches is of 599 μm (~48.5%) with respect to the "reduced number of patches" configuration and 2.072 mm with respect to the "original" configuration (~76.5%).

4.6.1.3 Stack-Up Variation: "Sandwich" Configuration
The third proposed configuration exploits the possibility of introducing two further layers in order to make the EBG solution even more compact than the previous ones. The new configuration, originally presented in Ref. [38], is named "sandwich" since the first row of patches is placed on the layer directly below the traces (as previously) and the other row of patches is placed on another layer with a reference solid layer in between. The patches on different layers are connected together by a via hole isolated by the intermediate reference plane by an antipad. Figure 4.67a and b shows both the top view and the stack-up of the EBG "sandwich" geometry. This last geometry is based on the same previous constraint of 127 μm for the bridge width. The optimal bridge length in order to match with a sufficient accuracy the design notch frequency of 8 GHz is achieved by an optimization procedure by using a the full-wave three-dimensional solver [18]. In the case of the EBG "sandwich" configuration, the optimized parameters consist of a single-side bridge length of 381 μm, accounting for the length of the path taken at the center of the bridge strip. The total length from patch to patch considering also the vertical via is 978 μm.

Figure 4.67c shows the behavior of the filter S_{cc21} transfer function comparing it with the other two miniaturization cases plus the "original" baseline case.

The filter effectiveness is reduced since the notch depth goes from −38.45 dB for the original EBG to −20.2 dB for the sandwich case; however, the notch level is still considered acceptable for many applications. The EBG sandwich configuration is quite effective in reducing the allocated filter surface despite the need of two additional stack-up layers. With this EBG configuration, the reduction of filter size is equal to 2.311 mm more than what obtained with the "meandered" version, whereas it is approximately 4.39 mm more compact than the "original" EBG solution. In practice, in terms of layout area, the required filter surface has been reduced respectively by 44.1% with respect to the "meandered" EBG, 49.8% with respect to the "reduced number of patches"

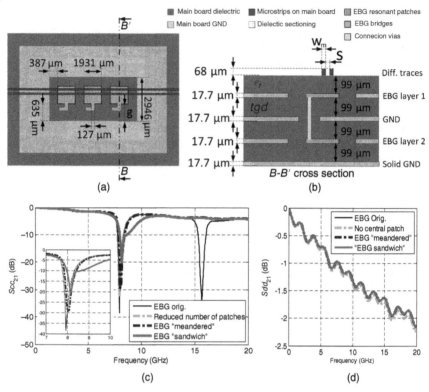

Figure 4.67 EBG-based CM filter "sandwich" configuration: (a) Top view of the first EBG layer. (b) Stack-up side view. Performance obtained compared with the other solutions: (c) $|S_{cc21}|$ responses. (d) $|S_{dd21}|$ responses.

EBG, and 59.9% with respect to the "original" EBG without significantly changing the common mode filtering performance.

Checking the differential mode insertion loss performance for the various configurations, the results obtained in all the miniaturized EBG-based CM filters have been comparatively analyzed in Figure 4.67d. All the solutions provide the same level of performance. The surface savings gathered with the adoption of the illustrated techniques are listed in Table 4.7.

4.6.1.4 Minimization of Powerplane Coupling and Cross Talk Features

In order to minimize the coupling effects between the EBG filter and the rest of the surrounding PCB structure, especially the cavities formed by the power/reference planes, we add vias around the EBG structure. These vias stitch the plane around the patches and the underneath continuous plane at the same potential. Figure 4.68 shows the top view of the stitched configuration for the

Table 4.7 Occupied rooms and surface reduction for the proposed three miniaturization techniques with respect to the original EBG case (baseline).

Model type	Frequency (GHz)	Notch depth (dB)	Occupied area (mm²)	% of surface reduction
Original EBG	7.94	−38.45	53.87	–
"Reduced number of patches"	7.94	−32.02	43.14	19.92
"Meandered"	8.02	−29.32	38.62	29.24
"Sandwich"	8.18	−20.2	21.64	59.83

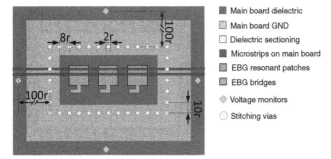

■ Main board dielectric
▨ Main board GND
☐ Dielectric sectioning
■ Microstrips on main board
▨ EBG resonant patches
▨ EBG bridges
◈ Voltage monitors
◯ Stitching vias

Figure 4.68 Top view of the EBG configuration named "sandwich" in presence of stitching vias.

"sandwich" configuration. The vias, of radius r, are placed at a distance of $\lambda/60$ to each other (equivalent to $10 \cdot r$) and at $\lambda/75$ (equivalent to $8 \cdot r$), where λ is the wavelength at the 8 GHz design frequency.

The application of the stitching to the "meandered" and "reduced number of patches" configurations give rise to configurations very similar to Figure 4.68, which are not reported here, for the sake of brevity, since this is an almost straightforward extension.

The presence of the vias have an impact on the S_{cc21} of the single differential pair considered for this analysis. For the three configurations, Figure 4.69 summarizes these results. The stitching improves the filtering performances especially for the sandwich configuration since a deeper notch occurs. For the other two configurations ("reduced number of patches" and "sandwich"), the via stitching generates a shift of the S_{cc21} notch with respect to the design frequency. This frequency shift can be compensated by slightly changing the dimensions of the bridge in order to adjust its inductance value.

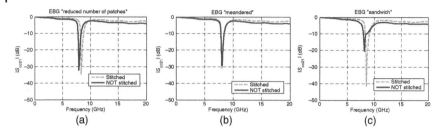

(a)   (b)   (c)

Figure 4.69 Comparison of the S_{cc21} with and without stitching vias for (a) "reduced number of patches," (b) "meandered," and (c) "sandwich" configurations.

To quantify the reduction of coupling from the EBG cavity and the power/reference planes, four voltage monitors have been placed in between the patterned plane and the reference plane in the positions indicated by diamonds at the four lateral edges of the board in Figure 4.68. They are at a distance of $100r$ from the edges of the gap between the plane and the patches. Figure 4.70 shows the magnitude of the voltage between the two planes without the vias (continuous line, "not stitched") and in presence of the stitching (dashed line, "stitched") for the three configurations.

A significant decrease in the coupling between the EBG cavity and the rest of the planes is achieved with the stitching vias. For all three configurations, an average reduction of the radiated field between the planes of approximately −40 to −50 dB is clearly visible.

Considering the cross talk effects, it can be shown that the coupling expressed by the values of FEXT and NEXT for all the three strategies does not highlight any clear superior performance of the three proposed configurations as reported in Ref. [38].

As further figure of merit, the normalized total radiated power (TRP) is introduced and defined as the magnitude of the TRP expressed in microwatt

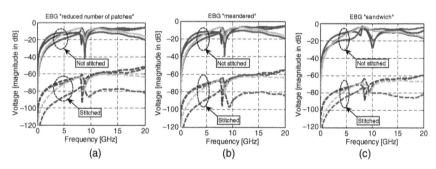

(a)   (b)   (c)

Figure 4.70 Magnitude of the voltage monitors between the planes for stitched and not stitched configurations. (a) Reduced number of patches. (b) Meandered. (c) Sandwich. (Reproduced from Ref. [38] with permission of IEEE.)

Table 4.8 Total radiated power for the three different miniaturization strategies.

EBG solution	TRP (μW)	Notch frequency (GHz)	Notch peak amplitude (dB)	Normalized TRP (μW/dB)
Reduced number of patches	920	7.94	−32.02	28.73
Reduced number of patches stitched	5483	8.36	−34.52	158.83
Meander	2600	8.02	−29.32	88.67
Meander stitched	2183	8.04	−30.93	70.57
Sandwich	280	8.18	−20.2	13.86
Sandwich stitched	1656	8.54	−41.42	39.98

divided by the depth of the notch at the notch frequency in decibel. The lack of a clear trend in the TRP data does not allow us to select the most efficient structure only from a radiation point of view. Table 4.8 summarizes some relevant results.

4.6.2 Design Hints for Bandwidth Enlargement

In some EBG filter applications, there is a need to enlarge the CM filter bandwidth for compensating possible mismatches between the nominal data rate of the signals and the actual data rate, or is due to the variation in the dielectric constant on the PCB. One way to enlarge the width of notch of the filter is to use multiple EBG sizes, as shown in Figure 4.71 and as proposed in Ref. [24]. This larger bandwidth approach helps to ensure that the error coming

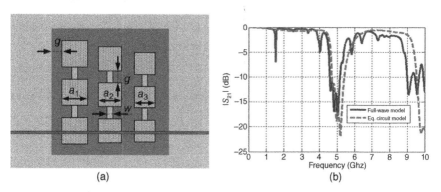

Figure 4.71 Multiple size EBG structures to widen filter bandwidth. (a) Top view geometry outline. (b) $|S_{21}|$ response. (Data from Ref. [24].)

from the design procedure (quantified in this case to be within 5%) still allows a reliable filter design.

The basic strategy for the bandwidth enlargement starts from the design of the three EBGs in Figure 4.71a as if they were designed employing the usual design steps. The basic resonating frequency was set at 5.4 GHz, and the obtained design parameters must be used for the definition of the central EBG ($a = a_2 = 2.55$ mm). Subsequently, the two lateral EBGs are designed using (4.4a)–(4.10), keeping constant the bridge geometry ($w = 0.4$ mm, $g = 1.3$ mm), but using $f_0^{new} = +5\%$ and $f_0^{new} = -5\%$ as design frequencies for each EBG, respectively. A larger variation, thus corresponding to a larger bandwidth, should be selected to take into account possible uncertainties in the dielectric constant value. A more reliable design can include more than three EBGs, although this increases the occupied layout area.

The notch frequencies for the EBG filter in Figure 4.71b will be such that a_1 has a resonant frequency at 5.15 GHz, a_2 has a resonant frequency at 5.4 GHz, and a_3 has a resonant frequency at 5.67 GHz. The computed EBG widths are $a_1 = 2.7$ mm, $a_2 = 2.55$ mm, $a_3 = 2.43$ mm. The complete filter response is obtained simulating both the equivalent circuit model and the full-wave model. The circuit model includes the EBG return paths computed by the cavity mode analytical procedure; the three different cavities have the following size: 13.93 mm $\times a_1$, 13.24 mm $\times a_2$, 12.61 mm $\times a_3$, Figure 4.71b shows the final results. The overall bandwidth has been increased as desired. The full-wave simulation results show a bandwidth of 600 MHz at -10 dB (12% fractional bandwidth) centered at 4.87 GHz (between 4.58 and 5.17 GHz). The equivalent circuit model was able to predict the filter bandwidth within acceptable limits, providing a 710 MHz bandwidth centered at 5.05 GHz.

4.7 EBG-Based CM Filters: Hardware Measurements

In this final section, we will show some hardware measurements that have been conducted with the aim of experimentally validating the EBG-based CM filter structures developed previously. In this section, the hardware setup will be fully illustrated and, subsequently, the correlation between computed and measured results will be accurately presented, accounting for both a frequency domain correlation and a time domain characterization.

4.7.1 Hardware Setup

Following the requirements of a practical case [36], the target frequency for the filters to build is set to 6.4 GHz. The stack-up available for manufacturing the board is based on a six-layer PCB, as shown in Figure 4.72.

Figure 4.72 Stack-up data of the manufactured test board, including the geometrical and technological details.

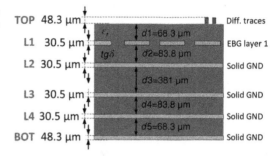

Only three layers are used for the filter design: top layer for the microstrip signal, layer L1 for the EBG, and layer L2 for the solid ground reference. The nominal dielectric constant is $\varepsilon_r = 3.5$ with a loss tangent factor $tg\delta = 0.007$. The computed equivalent length of the solid plane is $P = 12.528$ mm computed by (4.4b). The corresponding wavelength from (4.9) is $\lambda = 25.1$ mm. The number of patches is fixed at $M = 3$, yielding the optimum $a/\lambda = 0.105$ and the patch size $a = 2.631$ mm. The ratio $w/a = 0.125$ is used thereby setting $w = 0.329$ mm. The bridge length g can be evaluated using the bridge inductance from (4.8), leading to $g = 0.475$ mm. The designed EBG has an area of $A \times a = 23.25$ mm^2, where $A = M \cdot a + (M - 1) \cdot g$. It is important to highlight that compared to a solid patch-based filter, as in Ref. [2], which would require an area of 156.95 mm^2, the EBG-based filter solution shrinked the layout area by approximately 85%. In terms of S_{cc21} response, the responses from the equivalent circuit model together with the 3D full-wave model are compared in Figure 4.73a. Because of this, the EBG designs have been refined accounting for the shifts from the 6.4 GHz target frequency in both the full-wave model and the circuit model, the new results are shown in Figure 4.73b. Now the dimensions are $a = 2.46$ mm, $w = 0.307$ mm, $g = 0.454$ mm, for the full-wave model and $a = 2.51$ mm, $w = 0.314$ mm, $g = 0.46$ mm for the circuit model.

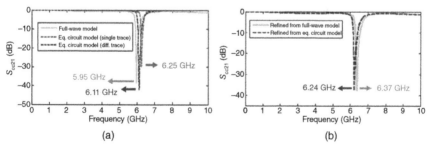

Figure 4.73 (a) Full-wave model compared to the single trace and differential trace equivalent circuit models. (b) Full-wave models of the filter refined starting from the initial full-wave model and the equivalent circuit model. (Reproduced with permission from Ref. [36].)

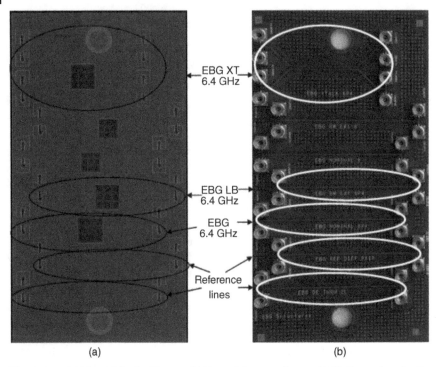

Figure 4.74 (a) View of the final layout. (b) View of the manufactured PCB. (Reproduced with permission from Ref. [36].)

4.7.2 Model-to-Hardware Correlation

4.7.2.1 Frequency Domain Measurements and Model Refinement

The real final PCB layout is depicted in Figure 4.74a, given by the dimensions listed in the previous section. A picture of the manufactured PCB is also shown in Figure 4.74b, which highlights different test sites for characterization, that is, a reference differential pair without EBG, a cross talk measurement site, and two filter test sites with different bandwidth.

The measurement of selected samples denoted slightly different values of dielectric and metal thickness from the nominal values, as shown in Figure 4.75a, which shows the actual size as well as the nominal values within parenthesis. An additional full-wave simulation was performed based on the measured stack-up. Results in Figure 4.75b show negligible difference in terms of the $|S_{cc21}|$ notch frequency, indicating that the typical PCB stack-up manufacturing variation does not have significant impact on the EBG resonant behavior. The measured stack-up size will be employed in all the following simulations.

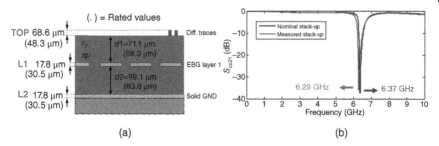

(a)   (b)

Figure 4.75 (a) Updated stack-up after cross section measurements; the values within parenthesis include the nominal values considered in the EBG design, as from Figure 4.74. (b) $|S_{cc21}|$ comparison of the model simulation based on the nominal and on the stack-up. (Reproduced with permission from Ref. [36].)

The measured patch size is slightly larger than the specification, which would shift the first EBG resonance to lower frequency. An opposite effect would be seen due to the larger bridge width w (from 0.31 to 0.32 mm). However, the overall effect of the simultaneous variation of a and w would have a minimal net impact on f_{TM10}.

It is possible to build an equivalent circuit model [31] of the EBG-based CM filter, like the one shown in Figure 4.44, taking into account the measured dimensions of the manufactured PCB with differential microstrip pair long around 5 cm.

The equivalent model contains the four SMP connectors as shown in Figure 4.57a. The measurement setup includes four 15 cm cables to match the SMP connectors at the PCB to the K connectors at the VNA side. These cables are modeled and cascaded with the full-wave simulation data for the EBG cavities. The cable model assumes a 15.2 cm length, a dielectric permittivity of $\varepsilon_r = 2$, and loss tangent $tg\delta = 0.005$. The manufactured PCB was also measured to extract the dielectric permittivity and loss tangent factor. The reference pair (the bottom-most link in Figure 4.74b) was measured in order to get the baseline behavior. The dielectric permittivity and the loss tangent factor that best fit the measurements are found to be $\varepsilon_r = 4.2$ and $tg\delta = 0.018$. These dielectric properties deviate significantly from the nominal design values used in the synthesis procedure ($\varepsilon_r = 3.5$ and $tg\delta = 0.007$), which is the major cause of the slight discrepancy between the measured and the modeled filter notch frequency.

4.7.2.2 Model-to-Hardware Correlation in the Frequency Domain

A preliminary comparison, in terms of S_{cc21} responses, has been done between the simulated filters (with nominal stack-up dimensions and with updated stack-up information obtained by the measurements) and the manufactured filters, and is shown in Figure 4.76.

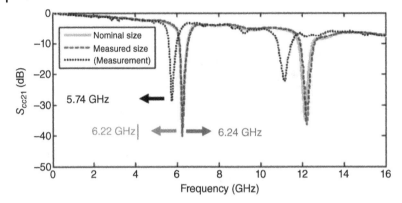

Figure 4.76 Comparison among the measured and simulated $|S_{cc21}|$. (Reproduced with permission from Ref. [36].)

From these results, we see that the size variations between nominal and actually measured geometry values do not significantly affect the results since the effects of larger a and w cancel each other. In addition, the results of the complete simulation model, taking into account also the SMP connectors, are consistent with the more simple original model developed for design purposes. In particular, the $|S_{cc21}|$ notch moves slightly from 6.29 to 6.22 GHz, as also shown in Figure 4.75b. From the previous results, it is noted that the variation of the electric permittivity has a greater impact on the filter performance, resulting in a 10% error in the filter notch frequency associated with the measured data (shift from 6.4 to 5.75 GHz). To account for this discrepancy between the computed results and the measurement, the simulation model was updated with the dielectric properties extracted from measurement ($\varepsilon_r = 4.2$, $tg\delta = 0.018$). The refined model-to-hardware correlation is shown in Figure 4.77 where a good match can be found for $|S_{cc21}|$. The measured and simulated differential mode and common mode insertion and return loss also agree well with each other.

The cross talk test site (EBG XT) shown at the top in Figure 4.74a consists of two adjacent differential microstrip pairs crossing the filter on the same EBG patch. The purpose of this test case is to investigate both the filter performance and the cross talk impact when the filter is applied to more than one trace pair. The measurements are performed using a four-port setup. The first two ports are connected at the left ports of the bottom pair, whereas the third and fourth ports are connected at the right side ports. During each measurement, the unused ports are terminated with 50 Ω loads. The results are shown in Figure 4.78. Basically, from a straightforward analysis of the results, both the CM NEXT and FEXT are relatively high and present a peak corresponding to the notch in the insertion loss. Therefore, the EBG cavity behaves as a coupling

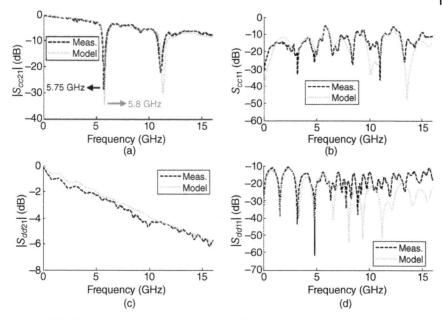

Figure 4.77 Final model to hardware correlation with model based on the measured dielectric properties ($\varepsilon_r = 4.2$, $tg\delta = 0.018$). (a) $|S_{cc21}|$. (b) $|S_{cc11}|$. (c) $|S_{dd21}|$. (d) $|S_{dd11}|$. (Reproduced with permission from Ref. [36].).

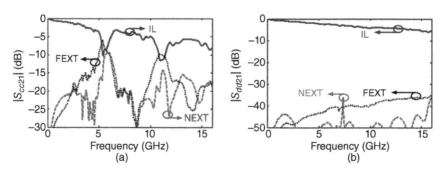

Figure 4.78 (a) Common mode and (b) differential mode insertion loss and cross talk. (Reproduced with permission from Ref. [36].)

path between two adjacent differential pairs, which is an undesirable effect that must be avoided. As mentioned in Section 4.4, regarding the real implementation issues of onboard EBG-based CM filters, this result will not encourage the layout of many pairs on the same filter. Also, the notch depth in $|S_{cc21}|$ is smaller ($\approx -8\,\text{dB}$) than that for the single pair filter, as shown in Figure 4.77a ($\approx -28\,\text{dB}$). The differential parameters, instead, can be considered satisfactory, with an S_{dd21} similar to the single pair case and a cross talk below $-35\,\text{dB}$.

4.7.2.3 Time Domain Measurements and Characterization

In order to complete the overall experimental description of the onboard EBG-based CM filters, it is required to investigate their performances in the time domain. The arrangement of the hardware setup for an oscilloscope is shown in Figure 4.79. It includes a differential PRBS/clock generator employed as the source, then the two single-ended signals pass through a phase shifter that provides the necessary time skew between the two signals for generating the common mode signal entering the EBG filter under test. Once the signals pass through the filter located on the test board, the two signals reach the oscilloscope. It is necessary to recall that due to the large variation of the dielectric permittivity that shifts the filter notch to 5.75 GHz, the input clock signal is set to 5.8 GHz in the subsequent time domain measurement.

The reference line (the bottom-most pair in Figure 4.74) is measured first, and the phase shifter is adjusted to minimize the common mode signal detected at the scope. The preliminary time domain measurements are shown in Figure 4.80. The minimum peak-to-peak common mode signal through the reference line is 5 mV. Next, the common mode signals with different amplitude are generated by tuning the phase shifter. Three cases are considered by adding 6, 14, and 21 ps time skew between the two single-ended signals, corresponding to 24.0, 47.5, 71.6 mV measured peak-to-peak common mode (Figure 4.81). Then the test setup is applied to the single notch EBG filter. The time domain results are shown in Figure 4.82. More than 70% CM reduction is achieved for the 5.8 GHz clock signal.

The cross talk case is also characterized in the time domain. Figure 4.83a illustrates the measurement setup. The third and fourth ports are moved from the bottom pair, for the insertion loss measurements, to the top pair for the FEXT measurements. The other four connectors are terminated with 50 Ω loads. The output common mode along the bottom pair is shown in Figure 4.83b, whereas the FEXT is shown in Figure 4.83c. The EBG filter reduces its effectiveness due to the smaller notch in the $|S_{cc21}|$, as shown in Figure 4.78a. The far-end cross talk is shown in Figure 4.83c. Although the differential cross talk is negligible ($\approx 5\,\text{mV}$ peak–peak), the common mode coupling to the adjacent pair has 24.5 mV amplitude. Therefore, the large common mode coupling through the EBG cavity at its resonance suggests that no more than one differential pair should be routed on the same EBG filter.

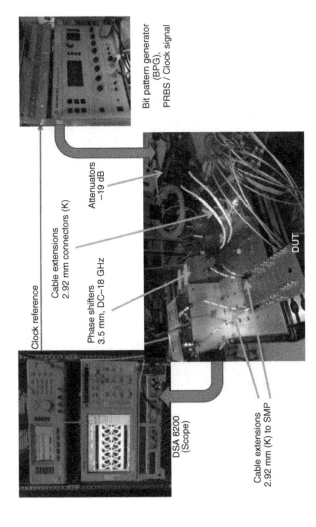

Figure 4.79 Time domain measurement setup including connectors.

Clock reference

Cable extensions
2.92 mm connectors (K)

Attenuators
−19 dB

Phase shifters
3.5 mm, DC−18 GHz

Bit pattern generator
(BPG),
PRBS / Clock signal

DSA 8200
(Scope)

Cable extensions
2.92 mm (K) to SMP

DUT

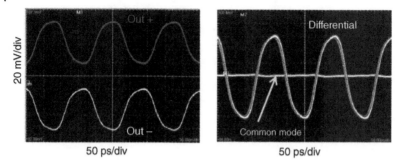

Figure 4.80 Initial test with minimized common mode (5 mV peak–peak). (Reproduced with permission from Ref. [36].)

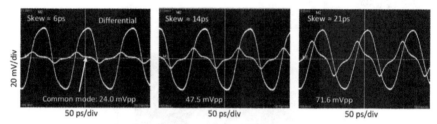

Figure 4.81 The three test cases characterized by different time skews and thus different input common mode amplitudes (reference through differential line). (Reproduced with permission from Ref. [36].)

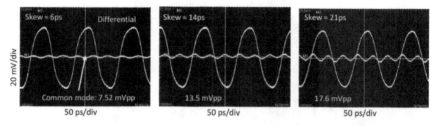

Figure 4.82 Measurement of the nominal EBG filter for the three input common mode amplitudes. (Reproduced with permission from Ref. [36].)

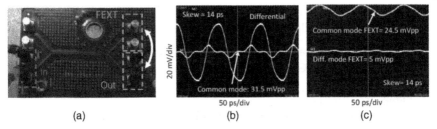

(a)   (b)   (c)

Figure 4.83 Measurements of the cross talk experiment (case of 14 ps time skew). (a) The measurement setup. (b) The common mode for the through pair increased to 31.5 from 13.5 mV in the case of EBG with one microstrip pair. (c) A large common mode coupling is found from the excited pair to the adjacent one; the FEXT peak–peak common mode is 24.5 mV. (Reproduced with permission from Ref. [36].)

4.8 Conclusions

It is interesting to underline the main advantages and drawbacks of using onboard EBG structures. While not fully comprehensive, we list the main benefits [39–49] seen from the following viewpoints: performance reached in terms of common mode and differential mode insertion loss, reliability of the proposed technology, implementation cost/effort, layout flexibility, and radiated power.

The following advantages are found from the viewpoint of improved CM filtering:

- The onboard EBG-based CM filter structures, both in their microstrip and stripline forms, present quite good differential mode insertion loss features together with very appealing common mode insertion loss features. In particular, the S_{dd21} behavior resulted in configurations that had low attenuation of the intentional differential signal, while the S_{cc21} behavior showed that the common mode signals are filtered by deep notch levels (deeper than −20 dB) over a quite wide range of frequencies.
- The filtering features are obtained through the use of a simple design workflow that is almost general, thus enabling a good scalability to different types of filters.

We note that the provided results seem to be controllable (from a manufacturing point of view) and most of the imperfections coming from the building process can be compensated for through the use of proper strategies. In fact, we have shown how some real manufacturing uncertainties can automatically compensate each other and, therefore, the resulting filtering performance is not heavily affected.

Concerning issues related to the PCB surface, which relate also to layout flexibility issues, the following advantages and disadvantages can be summarized:

- With respect to more classical simple patch-resonant structures, the employment of onboard EBG-based CM filters has clear advantages, since it has been shown a significant surface reduction using further miniaturization techniques (see Section 4.6).
- The building process of much reduced EBG solutions like the "meandered" and "sandwich" configurations requires the designer to make additional analysis after the standard design procedure has been applied. This could be a drawback since the refinement of such filters requires the employment of computationally intensive 3D solvers.
- From the viewpoint of implementation cost/effort, it can be easily concluded at the end of this chapter that no extra costs are required for the PCB designer and the EBG filter sizing effort is not unbearable if compared with the beneficial effects produced.

In summary, the onboard EBG-based CM filters technology is effectively appealing for CM noise reduction even though the discussed cross talk levels prevent their use with more than one trace over the same EBG. Also, some additional efforts still need to be spent for TRP reduction.

References

1 B.R. Archambeault, *PCB Design for Real-World EMI Control*, Kluwer Academic Publishers, Norwell, MT, 2002.
2 F. de Paulis, B. Archambeault, S. Connor, and A. Orlandi, Electromagnetic band gap structure for common mode filtering of high speed differential signals, in *Proc. of the IEC DesignCon 2011, Santa Clara, CA*, January 30–February 2, 2011.
3 D. Sievenpiper, L. Zhang, R.F.J. Broas, N.G. Alexopolous, and E. Yablonovitch, High-impedance electromagnetic surfaces with a forbidden frequency band. *IEEE Trans. Microw. Theory Tech.*, Vol. 47, No. 11, 1999, pp. 2059–2073.
4 J. Qin, O.M. Ramahi, and V. Granatstein, Novel planar electromagnetic bandgap structures for wideband noise suppression and EMI reduction in high speed circuits. *IEEE Trans. Electromagn. Compat.*, Vol. 49, No. 3, 2007, pp. 661–669.
5 T.L. Wu, C.C. Wang, Y.H. Lin, T.K. Wang, and G. Chang, A novel power plane with super-wideband elimination of ground bounce noise on high speed circuits. *IEEE Microw. Wirel. Compon. Lett.*, Vol. 15, No. 3, 2005, pp. 174–176.
6 S. Shahparnia and O.M. Ramahi, Electromagnetic interference (EMI) reduction from printed circuit boards (PCB) using electromagnetic bandgap

structures. *IEEE Trans. Electromagn. Compat.*, Vol. 46, No. 4, 2004, pp. 580–587.

7 T. Kamgaing and O.M. Ramahi, Design and modeling of high impedance electromagnetic surfaces for switching noise suppression in power planes. *IEEE Trans. Electromagn. Compat.*, Vol. 47, No. 3, 2005, pp. 479–489.

8 T.L. Wu, Y.H. Lin, T.K. Wang, C.C. Wang, and S.T. Chen, Electromagnetic bandgap power/ground planes for wideband suppression of ground bounce noise and radiated emission in high-speed circuits. *IEEE Trans. Microw. Theory Tech.*, Vol. 53, No. 9, 2005, pp. 2935–2942.

9 F. De Paulis, A. Orlandi, L. Raimondo, and G. Antonini, Fundamental mechanisms of coupling between planar electromagnetic bandgap structures and interconnects in high-speed digital circuits: Part I—microstrip lines, in *Proc. of the EMC Europe Workshop 2009, Athens, Greece*, June 11–12, 2009.

10 F. De Paulis and A. Orlandi, Signal integrity analysis of single-ended and differential striplines in presence of EBG planar structures. *IEEE Microw. Wirel. Compon. Lett.*, Vol. 19, No. 9, 2009, pp. 554–556.

11 J. Qin and O.M. Ramahi, Ultra-wideband mitigation of simultaneous switching noise using novel planar electromagnetic bandgap structures. *IEEE Microw. Wireless Compon. Lett.*, Vol. 16, No. 9, 2006, pp. 487–489.

12 S.S. Oh, J.M. Kim, J.H. Kwon, and J.G. Yook, Enhanced power plane with photonic band gap structures for wide band suppression of parallel plate resonances. *IEEE Int. Symp Antenna Propag.*, Vol. 2B, 2005, pp. 655–658.

13 T.K. Wang, C.C. Wang, S.T. Chen, Y.H. Lin, and T.L. Wu, A new frequency selective surface power plane with broad band rejection for simultaneous switching noise on high-speed printed circuit boards. *IEEE EMC Symposium 2005, Chicago, USA*, 2005, pp. 917–920.

14 T.L. Wu, C.C. Wang, Y.H. Lin, T.K. Wang, and G. Chang, A novel power plane with super-wideband elimination of ground bounce noise on high speed circuits. *IEEE Microw. Wirel. Compon. Lett.*, Vol. 15, No. 3, 2005, pp. 174–176.

15 F. de Paulis, A. Orlandi, L. Raimondo, B. Archambeault, and S. Connor, Common mode filtering performances of planar EBG structures, in *Proc. of IEEE International Symposium on Electromagnetic Compatibility, Austin, TX*, August 17–21, 2009, pp. 86–90.

16 F. de Paulis, L. Raimondo, D. Di Febo, B. Archambeault, S. Connor, and A. Orlandi, Experimental validation of common-mode filtering performances of planar electromagnetic band-gap structures, in *Proc. of the IEEE International Symposium on Electromagnetic Compatibility, Ft. Lauderdale, FL*, July 25–30, 2010.

17 F. de Paulis, L. Raimondo, S. Connor, B. Archambeault, and A. Orlandi, Design of a common mode filter by using planar electromagnetic band gap structures. *IEEE Trans. Adv. Packaging*, Vol. 33, No. 4, 2010.

18 Computer Simulation Technology , CST Studio Suite 2010, www.cst.com, 2010.

19 D.M. Pozar, *Microwave Engineering*, 3rd ed., John Wiley & Sons, Inc., 2005.

20 F. de Paulis, L. Raimondo, D. Di Febo, and A. Orlandi, Routing strategies for improving common mode filter performances in high speed digital differential interconnects, in *Proc. of the 2011 IEEE Workshop on Signal Propagation on Interconnects (SPI'11)*, May 8–11, 2011, Naples, Italy.

21 IEEE Standard P1597, Standard for Validation of Computational Electromagnetics Computer Modeling and Simulation—Part 1, 2008.

22 A.P. Duffy, A.J.M. Martin, A Orlandi, G Antonini, T.M. Benson, and M.S. Woolfson, Feature selective validation (FSV) for validation of computational electromagnetics (CEM). Part I—the FSV method. *IEEE Trans. Electromagn. Compat.*, Vol. 48, No. 3, 2006, pp. 449–459.

23 A. Orlandi, Feature selective validation (FSV) tool. Available at http://uaqemc.ing.univaq.it/uaqemc/FSV_Tool/

24 F. de Paulis, B. Archambeault, M.H. Nisanci, S. Connor, and A. Orlandi, Miniaturization of common mode filter based on EBG patch resonance, in *Proc. of the IEC DesignCon, Santa Clara, CA*, 2012.

25 M.H. Nisanci, F. De Paulis, A. Orlandi, B. Archambeault, and S. Connor, Optimum geometrical parameters for the EBG-based common mode filter design, in *Proc. of the 2012 IEEE Symposium on Electromagnetic Compatibility, Pittsburgh, PA*, August 5–10, 2012.

26 T. Weiland, A discretization method for the solution of Maxwell's equation for six component fields. *Electron. Commun.*, Vol. 31, 1977, p. 116.

27 EMS PLUS , EZPP User Manual. Available at www.ems-plus.com.

28 G.T. Lei, R.W. Techentin, P.R. Hayes, D.J. Schwab, and B.K. Gilbert, Wave model solution to the ground/power plane noise problem. *IEEE Trans. Instrum. Meas.*, Vol. 44, No. 2, 1995, pp 300–303.

29 B. Archambeault, A. Duffy, H. Sasse, X. Kai Li, M. Scase, M. Shafiullah, and A. Orlandi, Challenges in developing a multidimensional feature selective validation implementation, in *Proc. of the IEEE Symposium on Electromagnetic Compatibility, Fort Lauderdale, FL*, July 25–30, 2010.

30 A. Orlandi, A.P. Duffy, B. Archambeault, G. Antonini, D.E. Coleby, and S. Connor Feature selective validation (FSV) for validation of computational electromagnetics (CEM). Part II—assessment of FSV performance. *IEEE Trans. Electromagn. Compat.*, Vol. 48, No. 3, 2006, pp. 460–467.

31 Agilent , Advanced Design System, 2009. Available at www.agilent.com

32 C.-T. Lei, R.W. Techentin, and B.K. Gilbert, High-frequency characterization of power/ground-plane structures. *IEEE Trans. Microw. Theory Tech.*, Vol. 47, No. 5, 1999, pp. 562–569.

33 S. Huh, M. Swaminathan, and F. Muradali, Design, modeling, and characterization of embedded electromagnetic band gap (EBG) structure, in *IEEE 17th Conference Electrical Performance of Electronic Packaging, San Jose, CA*, October 27–29, 2008.

34 N. Sankaran, S. Huh, M. Swaminathan, and R. Tummala, Suppression of vertical coupling using electromagnetic band gap structures, in *Proc. of the IEEE 17th Conference on Electrical Performance of Electronic Packaging, San Jose, CA*, October 27–29, 2008, pp. 173–176.

35 F.D. Paulis, L. Raimondo, and A. Orlandi, Incorporating embedded planar electromagnetic band-gap structures within multilayer printed circuit boards. *IEEE Trans. Microw. Theory Tech.*, Vol. 58, No. 7, 2010, pp. 878–887.

36 X. Gu, R. Rimolo-Donadio, Y. Kwark, C. Baks, F. de Paulis, M.H. Nisanci, A. Orlandi, B. Archambeault, and S. Connor, Design and experimental validation of compact common mode filter based on EBG technology, in *Proc. of the IEC DesignCon 2013*, January 28–31, 2013, Santa Clara, CA.

37 F. de Paulis, M. Cracraft, D. Di Febo, H.M. Nisanci, S. Connor, B. Archambeault, and A. Orlandi, EBG-based common mode microstrip and stripline filters: experimental investigation of performances and crosstalk, *IEEE Trans. Electromagn. Compat.*, Vol. 57 No. 5, 2015, pp. 996–1004.

38 C. Olivieri, F. de Paulis, A. Orlandi, S. Connor, and B. Archambeault, Miniaturization approach for EBG-based common mode filter and interference analysis, in *Proc. of the IEEE International Symposium on EMC*, March 15–20, 2014, Santa Clara, CA.

39 F. de Paulis, M. Cracraft, C. Olivieri, S. Connor, A. Orlandi, and B. Archambeault, EBG-based common-mode stripline filters: experimental investigation on interlayer crosstalk. *IEEE Trans. Electromagn. Compat.*, Vol. 57 No. 6, 2015, pp. 1416–1424.

40 F. De Paulis, A. Orlandi, L. Raimondo, B. Archambeault, and S. Connor, Common mode filtering performances of planar EBG structures, in *Proc. of the 2009 IEEE Symposium on Electromagnetic Compatibility, Austin, TX*, August 17–21, 2009.

41 F. De Paulis, L. Raimondo, B. Archambeault, S. Connor, and A. Orlandi, Compact configuration of a planar EBG based CM filter and crosstalk analysis, in *Proc. of the 2011 IEEE Symposium on Electromagnetic Compatibility, Long Beach, CA*, August 14–19, 2011.

42 L. Raimondo, F. De Paulis, and A. Orlandi, A Simple and efficient design procedure for planar electromagnetic band-gap structures on printed circuit boards, in *IEEE Trans. Electromagn. Compat.*, Vol. 53, No. 2, 2011, pp. 482–490.

43 F. De Paulis, L. Raimondo, D. Di Febo, and A. Orlandi, Trace routing strategies for improving common mode filter performances in high speed digital differential interconnects, in *Proc. of the 15th IEEE Workshop on Signal Propagation on Interconnects, Naples, Italy*, May 8–11, 2011.

44 F. De Paulis, L. Raimondo, S. Connor, B. Archambeault, and A. Orlandi, Compact configuration for common mode filter design based on planar electromagnetic bandgap structures, in *IEEE Trans. Electromagn. Compat.* Vol. 54, No. 3, 2012, pp. 646–655.

45 B. Mohajer-Iravani and O.M. Ramahi, Suppression of EMI and electromagnetic noise in packages using embedded capacitance and miniaturized electromagnetic bandgap structures with high-*k* dielectrics. *IEEE Trans. Adv. Packaging*, Vol. 30, No. 4, 2007, pp. 776–788.

46 Y. Toyota, K. Iokibe, R. Koga, A.E. Engin, T.H. Kim, and M. Swaminathan, Miniaturization of electromagnetic bandgap (EBG) structures with high-permeability magnetic metal sheet, in *Proc. of the IEEE International Symposium on Electromagnetic Compatibility*,Honolulu, Hawaii, July 9–13, 2007, pp. 1–5.

47 Y. Toyota, A.E. Engin, T.H. Kim, M. Swaminathan, and S. Bhattacharya, Size reduction of electromagnetic bandgap (EBG) structures with new geometries and materials, in *Proc. of the Electronic Component and Technology Conference*, 2006.

48 S.H. Hall, and H.L. Heck, *Advanced Signal Integrity for High-Speed Digital Designs*, John Wiley & Sons, Inc., 2009.

49 C. Kodama, C. O'Daniel, J. Cook, F. De Paulis, M. Cracraft, S. Connor, A. Orlandi, and E. Wheeler, Mitigating the threat of crosstalk and unwanted radiation when using electromagnetic bandgap structures to suppress common mode signal propagation in PCB differential interconnects, in *Proc. of the Joint IEEE International Symposium on EMC and EMC Europe 2015, Dresden, Germany*, August 16–22, 2015.

5

Special Topics for EBG Filters

5.1 Introduction

Previous chapters have discussed how to design the EBG filter knowing the desired frequency to filter and the PCB dielectric parameters [1,2]. However, it was soon discovered that the nominal dielectric constant and even dimensions of the metal layers were not always the exact parameters. This resulted in the filter notch being at an incorrect frequency [3,4]. In addition, the frequency of the data signal sometimes shows slight variations, again, making the EBG filter not effective.

The depth of the notch is not as important as the notch's bandwidth (BW). Achieving a 5–10 dB deep filter performance across a wider frequency range is more important than the notch depth. This led to the necessity to find ways to widen the filter notch bandwidth.

5.2 Increased Bandwidth Filter: Multiple Size Patches

One way to widen the notch of the filter is to use multiple EBG sizes, as shown in Figure 5.1. This larger bandwidth design helps to ensure that the error coming from the synthesis procedure (quantified in this case to be within 5%) still allows a reliable filter design.

The three EBG sections in Figure 5.1 are designed, starting from the EBG geometry obtained with the refinement procedure resonating at 5.4 GHz, and placing it as the central EBG ($a = a_2 = 2.55$ GHz). The side EBGs are designed using the procedures in Chapter 4, keeping constant the bridge geometry ($w = 0.4$ mm, $g = 1.3$ mm), and using f_0^{new} ±5% as reference frequencies. A larger variation, thus corresponding to a larger bandwidth, should be selected to take into account possible uncertainties in the dielectric constant value. A more reliable design can include more than three EBGs, although this increases the occupied layout area.

Electromagnetic Bandgap (EBG) Structures: Common Mode Filters for High-Speed Digital Systems,
First Edition. Antonio Orlandi, Bruce Archambeault, Francesco De Paulis, and Samuel Connor.

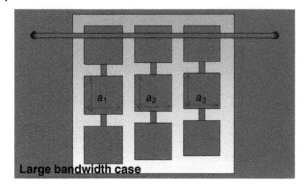

Large bandwidth case

Figure 5.1 Multiple size EBG structures to widen filter bandwidth.

Figure 5.2 shows a typical S_{cc21} performance of an EBG filter with all the patch sizes identical. Again, the depth of the notch ($\sim$4.5-5 GHz) is not as important as the notch width, which is pretty narrow in this example.

The notch frequencies for the EBG filter in Figure 5.3 will be such that a_1 has a resonant frequency of 5.15 GHz, $a_2 \to 5.4$ GHz, and $a_3 \to 5.67$ GHz The computed EBG widths are $a_1 = 2.7$ mm, $a_2 = 2.55$ mm, and $a_3 = 2.43$ mm. The complete filter response is obtained simulating both the equivalent circuit model and the full-wave model. The circuit model includes the EBG return paths computed by the cavity mode analytical procedure; the three different cavities have the following size: 13.93 mm $\times a_1$, 13.24 mm $\times a_2$, and 12.61 mm $\times a_3$, Figure 5.3 shows the final results. The overall bandwidth has been increased

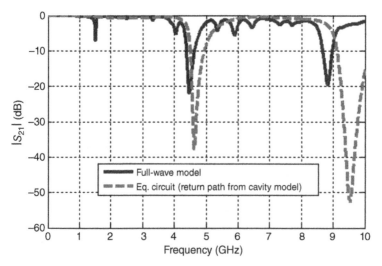

Figure 5.2 Typical EBG filter notch with all patch sizes identical.

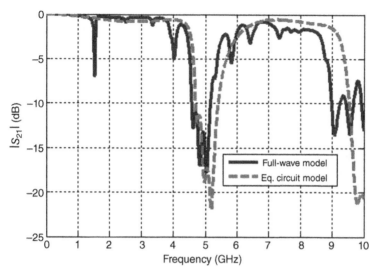

Figure 5.3 Insertion loss of the large bandwidth filter geometry.

as desired. The full-wave simulation results show a bandwidth of 600 MHz at −10 dB (12% fractional bandwidth) centered at 4.87 GHz (between 4.58 and 5.17 GHz). The equivalent circuit model was able to predict the filter bandwidth within acceptable limits, providing a 710 MHz bandwidth centered at 5.05 GHz. The FSV tool gives Grade = 5 and Spread = 5 for the ADM evaluation of the comparison in Figure 5.3.

5.3 Increased Bandwidth Filter: Multiple Size Bridge Width

This section deals with the enlargement of the filter bandwidth based on the use of EBGs resonating at different frequencies. This is achieved by following the procedure detailed in Section 6.4. The left and right EBG cavities, formed by the patches connected by bridges, are designed at −10% and +10% with respect to the 8 GHz target frequency, respectively. In the design, the patch size a and the bridge length g are kept constant, whereas the bridge width w of the −10% EBG vertical column and +10% EBG vertical column are recalculated as 2.87 and 8.42 mils, respectively. The model is shown in Figure 5.4, and the simulation results are given in Figure 5.5. The differential insertion loss does not change, whereas the common mode insertion loss presents three closely spaced notches, as expected. However, the depth of the notches is drastically reduced, providing about −3 dB within a 1.4 GHz bandwidth, compared to the −10 dB within 100 MHz bandwidth in the original case.

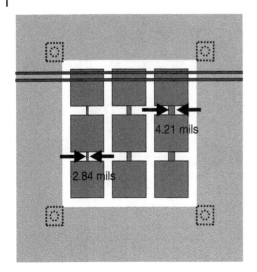

Figure 5.4 Multiple size EBG bridge widths to widen filter bandwidth.

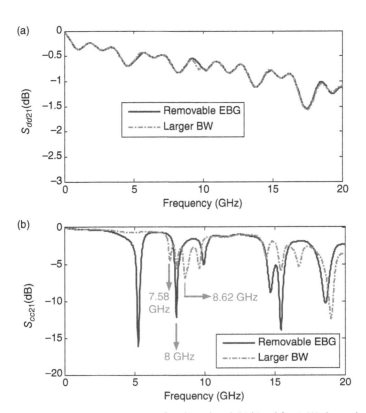

Figure 5.5 Frequency spectra of (a) $|S_{dd21}|$ and (b) $|S_{cc21}|$ for 8 GHz larger bandwidth filter.

5.4 Conclusions

Two straightforward techniques are used to increase the filter notch bandwidth. The depth of the notch is lower as the bandwidth is increased, but often only a small amount of filtering is required; so this is typically not an issue. However, if both a wide bandwidth and an increased filtering are required, then additional section of the EBG filter can be added.

There are many parameters that can be modified to change the filter bandwidth. While only two are shown here (patch size and bridge width), other parameters can also be modified.

References

1 F. De Paulis, A. Orlandi, L. Raimondo, B. Archambeault, and S. Connor, Common mode filtering performances of planar EBG structures, in *Proc. of the 2009 IEEE Symposium on Electromagnetic Compatibility, Austin, TX*, August 17–21, 2009.

2 F. De Paulis, L. Raimondo, S. Connor, B. Archambeault, and A. Orlandi, Design of a common mode filter by using planar electromagnetic bandgap structures, in *IEEE Trans. Adv. Packaging*, Vol. 33, No. 44, 2010, pp. 994–1002.

3 X. Gu, R. Rimolo-Donadio, Y. Kwark, C. Baks, F. de Paulis, M.H. Nisanci, A. Orlandi, B. Archambeault, and S. Connor, Design and experimental validation of compact common mode filter based on EBG technology, in *Proc. of the IEC DesignCon 2013*, Santa Clara, CA, January 28–31, 2013.

4 F. de Paulis, M.H. Nisanci, A. Orlandi, X. Gu, R. Rimolo-Donadio, Y. Kwark, C. Baks, B. Archambeault, and S. Connor, Experimental validation of an 8 GHz EBG based common mode filter and impact on manufacturing uncertainties, in *Proc. of the IEEE International Symposium on EMC*, Denver, CO, August 5–9, 2013.

6

Removable EBG Common Mode Filters

with contributions of Carlo Olivieri

Subsequent to the introduction of the onboard EGB-based common mode filters [1–14], this chapter describes their evolution to *removable EBG-based common mode* (R-EBG CM) filters [15,16]. The key feature of this kind of device is that PCB designers can replace an EBG-based filter with a different filter with a different frequency range without changing the layout of the main PCB. This chapter will also describe the different topologies available for the removable EBG filters, ranging from the configuration with the traces kept on the main PCB to the one with the differential pair transitioning into the removable EBG. Several details will be given on the increased performance of the removable structure using different substrate materials (i.e., organic, ceramic, etc. substrates), different cavity resonator geometries (EBG cavity versus rectangular cavities) and the design strategies specifically addressed to increase the filter band-stop bandwidth.

6.1 Design Concept of Removable EBG Filter

In many practical applications, it is important and required that the common mode filtering capabilities [15] should be associated with a specific, stand-alone component instead of being laid on the main PCB.

In order to be compliant with such a requirement, the evolution of the EBG-based CM filters technology has been driven toward an innovative, from a design viewpoint, concept named "removable EBG" concept [15,16].

The basic idea underlying this configuration is the possibility of having an EBG-based CM filter that is independent of the main PCB and its stack-up. This component has its own specific stack-up, and in general different from the stack-up used to fabricate the main PCB. Although other similar concepts exists in literature [17], the aforementioned design choice, that is, the elimination of

Electromagnetic Bandgap (EBG) Structures: Common Mode Filters for High-Speed Digital Systems,
First Edition. Antonio Orlandi, Bruce Archambeault, Francesco De Paulis, and Samuel Connor.
© 2017 by The Institute of Electrical and Electronics Eingineers, Inc. Published 2017 by John Wiley & Sons, Inc.

the EBG filter from the stack-up of the main PCB and transferring it to a stand-alone part, enables the electronic system designer to reach several goals that cannot be realized otherwise.

To list some of the improvements offered by the removable EBG structure with respect to the onboard EBG described in Section 4.1 (a planar EBG-based CM filter built inside the main PCB), the following points should be considered:

1) Layout design flexibility
2) CM filters and overall PCB design independency
3) Scalability of the overall design

The first two points show that the R-EBG solution allows the design of the main PCB with more flexibility. This means that the layout specialist can work in parallel with the system designer, since the filtering stages can be sized independent of the actual layout of the main PCB under development. This allows decoupling of the two aspects: main PCB layout and EMI filtering.

It is worth noting that the R-EBG allows a better use (for trace routing purposes) of the space underneath the component. In the onboard config-uration, as mentioned in Section 4.1, parts of some main PCB inner layers are occupied by the EBG layout. In the R-EBG, this does not happen and there is more allowable area on the main PCB for routing [18]. Also, the R-EBG configuration does not generate electromagnetic radiation among the main PCB layers as the onboard configuration does. On the other hand, since the R-EBG structure is external, the electromagnetic radiation tends to propagate in the surrounding space, possibly creating another kind of EMC problem.

As final consideration, it is also worth noting that the use of R-EBG allows one to change the filtering performance of the main PCB without having to modify the PCB structure and stack-up, but simply upgrading the R-EBG filters employed in the project (Figure 6.1).

6.2 Categorization of Filters and Structures

The implementations of the R-EBG, they can be classified into two main groups:

1) With respect to the position of the signal traces:
   a) Removable EBG filters lying on microstrips on the main PCB
   b) Removable EBG filters with traces entering inside the filter itself
2) With respect to the resonant properties of the structure geometry:
   a) Standard EBG cavity, using a grid of square patches
   b) Rectangular cavities, using solid rectangular patches.

In order to explain exactly the meaning of the aforementioned classifications, please note that the position and the structure of the signal traces approaching

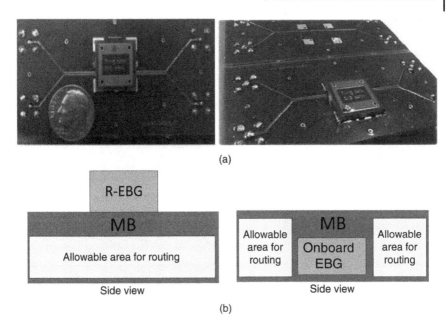

(a)

(b)

Figure 6.1 (a) Overall structure for a removable EBG filter. (b) Depiction of some of the beneficial effects gained, such as the possibility to have more allowable area for routing on the main PCB.

the R-EBG filter will affect, in further layout developments, the possibility of attachment of the filtering device to the main PCB of interest. In fact, if in the design process there are some signal traces running on the main PCB in the form of microstrips, the R-EBG structure should be selected in order to comply with the other layout constraints. These design choices may be mainly linked to the manufacturing requirements. On the other hand, excluding all the layout design requirements, the selection (and design) of the appropriate R-EBG filter would be linked to the specification of particular filtering properties desired. This is a design choice occurring in the development stage of the filter itself, which is usually the responsibility of the filter designer.

It must be mentioned that the implementation of 1(a) requires the generation of a void on the main PCB's metal layers below the R-EBG footprint. This layout requirement will be explained with more details in the following section, and it limits the layout flexibility on the main PCB often requiring solutions like those in 1(b), as shown in Figure 6.2b).

The internal structure of the resonant cavities that the EBG-based CM filter is based upon will affect the performance of the filters and are strongly dependent on the geometry that the designer will select for the desired application. In general, there exists a very wide set of possibilities for the geometry of an EBG

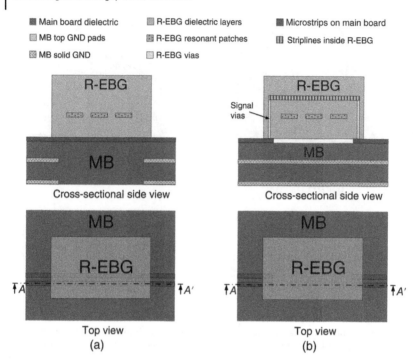

■ Main board dielectric   ▨ R-EBG dielectric layers   ■ Microstrips on main board

▨ MB top GND pads   ▨ R-EBG resonant patches   ▥ Striplines inside R-EBG

▨ MB solid GND   ▢ R-EBG vias

Figure 6.2 Different signal trace structures that can be used for a removable EBG. (a) Design solution with the traces kept on the main PCB. (b) Solution with the differential pair moved inside the removable part as a stripline pair. Some details such as reference pads and other metal layers have been hidden for the sake of simplicity.

resonant cavity, spanning from the more classical array of square patches (which can be linked each other by the presence of conductive connections called a bridge) until reaching more specific geometries like the one accounting for a set of long rectangular solid patches. Each different kind of resonant cavity can be in turn tailored to achieve specifically designed frequency responses of the filter, for example, the best geometry to get the widest band-stop filter as possible. Without loss of generality, one can refer to the structures illustrated in Figure 6.3.

6.3 Removable EBG Common Mode Filters Design Approach

In order to explain which is best design approach for a removable EBG-based CM filter, it is necessary to recall that in the case of EBG CM filters on the main PCB, once the EBG cavity is designed along the return path of a

■ Main board dielectric   ▨ R-EBG dielectric layers

■ Microstrips on main board   ■ R-EBG internal rings

▢ MB top GND pads   ▨ R-EBG resonant patches

▨ MB solid GND   ▨ R-EBG bridges

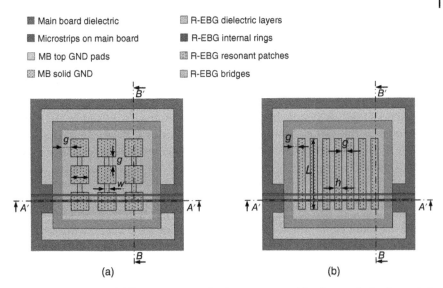

(a)   (b)

Figure 6.3 Top view of different structures for the resonant cavities that can be implemented for a removable EBG filter. (a) Design solution with a standard EBG patch grid. (b) Configuration with rectangular solid patches.

differential trace, the currents associated with the differential mode remain unaffected, while the common mode current flowing on the reference solid layer below the traces see the return path discontinuity created by the EBG. In this case, the common mode return current is forced to flow on the EBG patches and in the closest region of the solid plane between two EBG patches. As reported in Section 4.1, the mechanism enabling the reduction of the common mode current consists essentially in the energy transfer, between the common mode and the cavity mode, occurring when the displacement current flowing from the EBG to the solid plane below is able to excite the EBG resonance.

Even though the structure of a removable EBG filter is quite different from the case with the EGB filter on the main PCB (and much more complicated in some sense), the fundamental mechanism remains still the same. In addition to the key features of the EBG technology, previously introduced, which make these filters attractive, the use of standard multilayer PCB technologies, the simplicity of the design procedure, the reduced cost, and so on are still valid. To determine the filtering capabilities of the R-EBG structures, note that the electromagnetic behavior of the filter remains unchanged, with the common mode currents of the differential pair being responsible for the common mode to EBG cavity mode coupling.

In particular, recalling from the previous section and referring to the case of an R-EBG filter configuration accounting for signal traces running on the top of the main PCB as microstrips (see Figure 6.2a), the PCB area directly below the EBG is left void since this will require the return current to flow upward into the resonant cavities once it reaches the EBG footprint. This is done through vias connecting the reference GND plane of the main PCB to a metal ring laid out around the EBG, as illustrated in Figure 6.4. In this case, the width of the ring is set equal to the gap *g*. Also note that in some cases, these vias can also be connected to the top metal layer of the R-EBG in order to extend the current return path.

As depicted in Figure 6.4, the return current flows back and forth between the EBG layer and the top GND reference layer of the removable EBG, similar to what happens for the EBG-based CM filter on the main PCB. As can be seen from the same figure, the forward current outside the EBG area flows on the bottom side of the traces facing the associated return layer (i.e., the main PCB ground-reference), whereas it flows on the upper side when it enters the EBG, since the return path is above the traces.

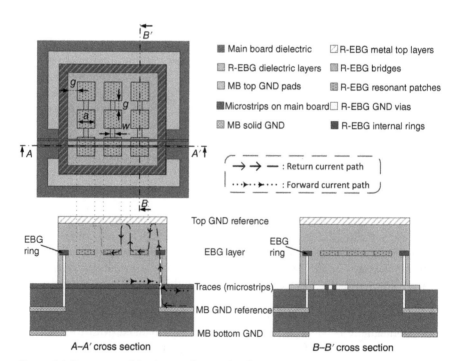

Figure 6.4 Structure and fundamental principle of the R-EBG CM filter; the scheme also reports the internal details of the R-EBG CM filter such as the metal ring around the EBG.

6.3.1 Design Approach and Guidelines

In order to explore the major topics of the design process for a removable EBG-based CM filter, the R-EBG structure illustrated in Figure 6.5 should be considered as a starting point for a typical design process. This structure, following the categorization previously introduced, belongs to a design case accounting for a filter with traces lying onto the main PCB and with an EBG cavity formed by a 3×3 grid of square patches. This could be considered as a good example for the design process since it responds, at least for the building of the resonant EBG cavities, to the guidelines for the onboard EBG structures [4–6,9,11] already explained in detail in Chapter 3 and Section 4.1.

As a first design step, once the target notch filter frequency has been set (in this case, equal to 8 GHz), the designer can preliminarily size the geometry of the resonant EBG cavity, in particular following the workflow illustrated in

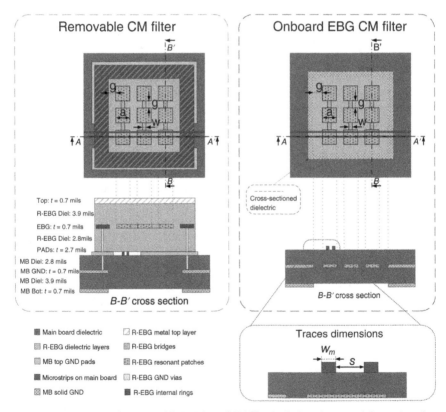

Figure 6.5 Structure of a removable EBG-based CM filter including the typical dimensions for a target frequency of 8 GHz. The design accounts for traces running on the main PCB and for a 3×3 grid of square patches as the EBG cavity.

Chapter 3 and in Refs [11,13]. The execution of this initial step, taking into account the constraints of the stack-up to be employed (in this case, the relative dielectric permittivity was chosen as $\epsilon_r = 3.6$ and the thicknesses of the metal layers are the ones indicated in Figure 6.5), will result in the following geometrical dimensions: patch size $a = 76.06$ mils, bridge width $w = 9.5$ mils, and bridge length (equal to the gap between patches) $g = 15.26$ mils, and ring width $= 2g$.

As can be concluded from this preliminary step of the design process, the removable version of the EBG-based CM filter has the same top view as the onboard case: This is a straightforward consequence of the fact that the underlying physical principles are similar for both types of filters.

In addition, it must be noted that in this specific case, the differential microstrip on the top of the main PCB was originally designed for 85 Ω differential impedance, with the trace width $w_m = 5$ mils and the trace separation $s = 5.5$ mils. Actually, this is not a conservative choice since most of the modern high-speed systems must be compliant with the well-known PCIe 3 Standard [19], which makes use of such a specification.

Even though the preliminary design step of an R-EBG structure is not different from what is already done for the main PCB case, it should be noted that when designing a removable EBG-based CM filter, there are many other aspects that will be involved that can make the design process more challenging than its planar counterpart.

For instance, it is of great interest to study the frequency responses of the two filters just introduced. In particular, referring to the main PCB version as the "original" model and computing the common mode and the differential mode insertion loss parameters for both designs, we observe that there are several differences between them. All these aspects require a more detailed explanation and will be covered in the following discussion.

The two aforementioned filters, the original and the removable EBG, are studied by the use of the full-wave, three-dimensional, finite integration technique (FIT)-based electromagnetic solver CST MicroWave Studio® [20]. Both filters are properly implemented in the selected modeling environment.

The two key figures of merit are the differential and common mode insertion losses that are quantified by the two mixed-mode scattering parameters S_{dd21} and S_{cc21}, and are shown comparatively in Figure 6.6.

Note that S_{dd21} is only slightly affected by the removable filter case. The voided reference planes beneath the microstrip produce a discontinuity for both the common and differential modes. The differential mode return current flows mainly on the traces, although the voids and filter itself represent an additional discontinuity along the differential path leading to a change in the differential impedance. This consideration is supported by the fact that, when analyzing the differential impedance of the two filters, the removable solution shows a change from 85 to 78 Ω, while the "original" filter instead shows a discontinuity in the

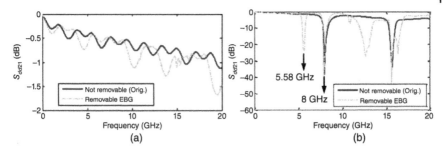

Figure 6.6 Comparison of (a) S_{dd21} and (b) S_{cc21} of the onboard/original EGB versus the removable EBG. (Reproduced from Ref. [16] with permission of IEEE.)

opposite direction: The impedance rises up to about 90 Ω as a consequence of an increment of distributed inductance. The two aforementioned behaviors are shown in Figure 6.7a. In order to better explain this fact, it is worth noting that for the R-EBG solution (with the voided reference planes below the traces), the connection of the removable part inherently provides a lower impedance return path for the common mode return currents that forces them to flow through the removable EBG filter.

Observing the main common mode filtering properties, a quick analysis of the S_{cc21} parameter shows the same design notch located at 8 GHz for both the original and removable filters, meeting the previously defined design requirements. However, another interesting thing is that an additional notch appears at a lower frequency (approximately 5.58 GHz) for the case of the R-EBG. This additional notch is a consequence of the much more complex structure of the filter itself, which includes the resonances of the EBG patches and also of several other resonating structures involved by the filter geometry.

An investigation of the spatial distribution of the magnitude of the electric field E explains this behavior. Figure 6.8 shows the spatial distribution of E at the

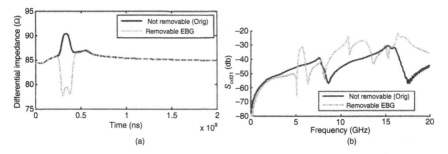

Figure 6.7 (a) Comparison of the differential impedance (evaluated by TDR) in the case of original EBG filter versus removable EBG-based CM filter. (b) Differential-to-common mode conversion scattering parameter S_{cd21} for onboard/original and R-EBG. (Reproduced from Ref. [16] with permission of IEEE.)

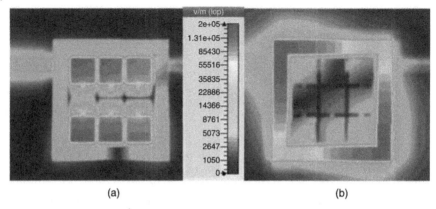

(a)   (b)

Figure 6.8 Spatial distribution of the electric field inside the removable EBG-based CM filter when it is excited by common mode signals. As shown by the two contour plots, there are different resonances at different frequencies. (a) E at 8 GHz. (b) E at 5.58 GHz. (Reproduced from Ref. [15] with permission of IEEE.)

first two resonant frequencies (5.58 and 8 GHz) on a longitudinal cutting plane between the top layer and the layer including the EBG patches. In both cases, the traces are directly excited by a common mode signal generated by a proper placing of the excitation ports.

The spatial distribution of E in Figure 6.8a is associated with the first EBG resonant mode at 8 GHz. It is worth noting that a larger electric field is present around the traces crossing the EBG (on the top left of the EBG), whereas a smaller electric field occurs on the right side due to the common mode electromagnetic energy filtered by the EBG. The remaining energy is propagated on the traces exiting the filter area. On the other hand, Figure 6.8b shows the resonant pattern at 5.8 GHz due to the resonance of the ring that surrounds the EBG. This resonance is excited by the vertical current, shown in Figure 6.4, flowing from the ring on the EBG layer toward the top layer of the removable EBG filter. Moreover, the additional notch in S_{cc21} is due to the ring, at 5.58 GHz, and others at higher frequency generate corresponding peaks in the mode conversion, as shown by the S_{cd21} in Figure 6.7b. The additional notches make the filter more effective because they enlarge the filter BW; however, the level of the mode conversion should be accurately quantified for an effective filter design.

These preliminary results validate the proposed filter design. However, in order to also test the scalability and reliability of this design approach, some changes can be applied to the developed model in order to show the flexibility reached by the R-EBG geometry.

Let us suppose that the implemented removable EBG-based CM filter has to be redesigned for the same target frequency, but the dielectric material included in the stack-up must be changed. Actually, this could be quite a reasonable scenario since the variation in the material can be due to the need of reducing the dimensions of the entire device or, if this is the case, forced by market and/or technological options.

The key features of the removable EBG-based CM filter is that the dielectric material being changed involves only the removable component and not the main PCB which has an independent stack-up.

Let us assume, without loss of generality, that the new material is characterized by a relative dielectric constant four times higher than the preceding one, meaning that this should allow a reduction of the overall filter size by approximately one half. At this point, the presented design procedure is applied again and the following dimensions are found: patch size $a = 40.85$ mils, bridge width $w = 5.1$ mils, and bridge length $g = 10.67$ mils. Similar to the previous example, the ring width can be set to $2g$. It is not difficult to verify that the layout of the R-EBG filter does not change significantly and that the resonant patch layer is practically only a rescaled version of the preceding one, according to the new dimensions. However, the removable PCB area savings is only 32% of the previous design.

The structure of the new R-EBG and the S-parameter results are shown in Figure 6.9 due to the shrinking of the dimensions. The key point is that the filter can be modified in its layout and material without changing the stack-up and substrate material of the main board. The S_{dd21} shows a limited variation between the case with $\epsilon_r = 3.6$ and $\epsilon_r = 14.4$. The common mode insertion loss (S_{cc11}), on the other hand, shows a reduction in the notch depth at 8 GHz (from -38 to -12 dB) and a shift in the notch due to the ring resonance (from 5.7 to 5.26 GHz).

The notch reduction is attributed to the weaker excitation of the resonance of the EBG cavity. It is caused by the size a of the EBG patch compared to the trace width ($2 \cdot w_m + s$). The ratio is 0.2 in the original case ($\epsilon_r = 3.6$), whereas it reaches 0.38 in the case of $\epsilon_r = 14.4$. Because of this, the filter size cannot be shrunk arbitrarily, but the EBG patch should be carefully sized to achieve an effective notch depth.

For the R-EBG filter with $\epsilon_r = 14.4$, a specific analysis was carried out to investigate the effect of the voided reference layers in the PCB below the EBG.

For this reason, a new simulation model was created. In this model, the metal layers of the main board have been replaced with solid metal one. The results in terms of S_{cc21} and S_{dd21} are shown in Figure 6.9 in comparison to the models with voided reference layers beneath the EBG. The main effect of including the PCB solid reference is that the design notch is reduced in its depth and shifted from 8 to 8.42 GHz.

The layout with solid reference planes on the main board allows the return current to keep flowing on the solid reference on the main board. In this case, less electromagnetic energy is diverted toward the EBG with respect to the case where the reference plane is voided. From a layout viewpoint, this situation

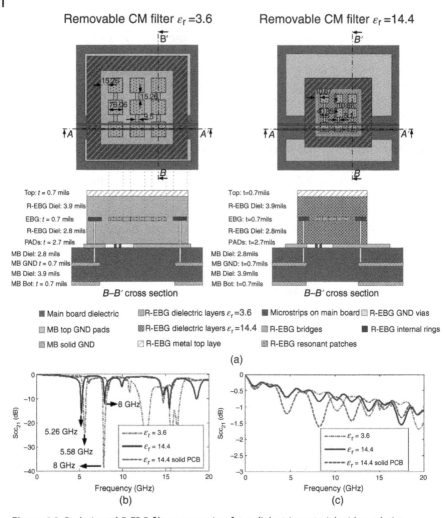

Figure 6.9 Redesigned R-EBG filter accounting for a dielectric material with a relative dielectric constant of $\varepsilon_r = 14.4$. Comparison of (a) R-EBG layout and cross section for two different dielectric materials. (b) S_{cc21}. (c) S_{dd21}. (Reproduced from Ref. [15] with permission of IEEE.)

limits the use of the R-EBG, since the filter is more effective only when the PCB layout allows the return current to properly excite the EBG cavity, implicitly requiring the presence of the voids on the main board generating, in turn, potentially less routing freedom on other layers of the main board.

This resonant behavior is confirmed by additional simulation models to investigate the effect of distance between the differential traces and the PCB

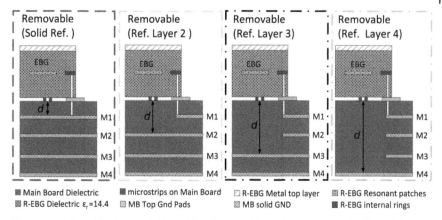

Figure 6.10 Analysis of the variation of the distance *d* from the traces and the first solid plane underneath.

solid layer, the models involved are briefly described in Figure 6.10. The S_{cc21} simulation results from these PCB main board configurations are shown in Figure 6.11.

As shown in Figure 6.11a and b, it can be observed that a significant reduction in the depth of the R-EBG filter notch and also a shift of the resonant frequency occur when the distance from the EBG layer to the nearest solid plane on the main board increases.

6.3.2 Removable EBG CM Filter Performances

6.3.2.1 Notch Bandwidth

As demonstrated in the last section, the features characterizing a removable EBG filter are similar to those related to the onboard version, for a similar

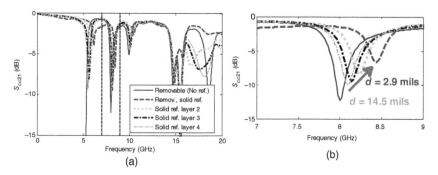

Figure 6.11 S_{cc21} for the R-EBG CM filter configurations in Figure 6.10. (a) S_{cc21} and (b) zoom between 7 and 9 GHz. (Reproduced from Ref. [16] with permission of IEEE.)

geometry of the EBG patches. Such similarities are also expected for the R-EBG filter bandwidth (BW). Previous chapters have shown that the exact notch frequency is dependent on the dielectric constant (and other terms). Since the exact dielectric constant is not always known, a wider notch bandwidth helps ensure the signal to be filtered is contained within the bandwidth of the notch filter.

According to the workflow introduced in Refs [13,16,21], the EBG cavities formed by the left- and right-most columns of patches are designed for a resonant frequency of −10 and +10% of the nominal frequency (8 GHz in this example). In order to maintain the overall external dimensions of the R-EBG CM filter, the patch size a and bridge length g are kept constant. However, the bridge width w of the left EBG vertical column (designed for a frequency −10% of the nominal frequency) and the right EBG vertical column (designed for a frequency +10% of the nominal frequency) are recalculated as 2.87 and 8.42 mils, respectively.

Therefore, the EBG layer geometry is affected only by a slight modification, producing a structure with a larger bandwidth (LBW) similar to the one shown in Figure 6.12. All the other dimensions remain unchanged.

The results shown in Figure 6.13 show that the differential mode insertion loss does not change, whereas the common mode insertion loss presents three closely spaced notches at frequencies $f_n - 10\% f_n$, f_n, and $f_n + 10\% f_n$, where f_n is the nominal or design frequency. The depth of each single notch is drastically reduced, providing about −3 dB within a 1.4 GHz BW, compared to the −10 dB with 100 MHz BW for the regular case in Figure 6.12a. This means that the strength of the original filtering action has been split over a greater number of resonant frequencies, resulting in the overall bandwidth larger than the original

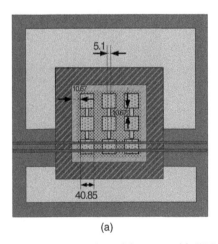
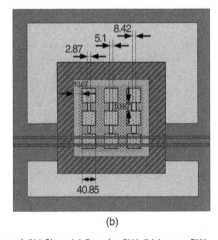

(a) (b)

Figure 6.12 Top view of the removable EBG-based CM filter. (a) Regular BW. (b) Larger BW.

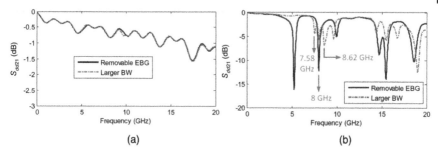

Figure 6.13 Frequency spectra of (a) $|S_{dd21}|$ and (b) $|S_{cc21}|$ for the removable EBG-based CM filter for larger BW in Figure 6.12b). (Reproduced from Ref. [15] with permission of IEEE.)

bandwidth but each single resonance is characterized by a smaller depth of the notch since the global common mode energy coupled to the EBG cavities does not change significantly with respect the case of only one resonant frequency f_n.

6.3.2.2 The Depth of the Notch

Another very relevant aspect to be addressed when designing a R-EBG CM filter is the depth of the notch at the design frequency, since this represents the effectiveness of the device for CM rejection purposes. The likelihood that the frequency to be filtered exactly matches the deepest part of the notch is low, making the depth important so that the depth of the wider part of the notch (the notch bandwidth) is deep enough for the amount of filtering desired. For sake of completeness, it must be said that there exist several parameters that will affect the depth of the notch. Most of them are associated with the geometry of the EBG resonant cavities, while others are associated with the materials.

Among the various possible parameters, the impact of the number of patches forming the removable EBG geometry plays a significant role and will be presented in this chapter. The original case with an array made by three columns of three EBG patches is compared with the case made by four columns of three patches (named 4EBGs) as in Figure 6.14a), and with another case with three columns of three patches (named 3EBGs) but having the ring length equal to the length of the four columns case, as in Figure 6.14b). Once again the ring width is set equal to 2g.

The comparisons of $|S_{dd21}|$ and $|S_{cc21}|$ are given in Figure 6.15. The four-EBG case provides a better filtering effect since the notch at 8 GHz reaches −20 dB. The larger ring in the three-EBG case increases the notch depth from −12 to −14 dB. The frequency value of the first notch moves down from 5.26 to 4.62 GHz for both cases. The ring resonance provides the additional notch at lower frequency with respect to the target nominal or design frequency (8 GHz in this example) and extends the effectiveness of the filters to other frequency components different from the one it has been mainly designed. Until now no direct relationship between the ring length and the ring resonant frequency has

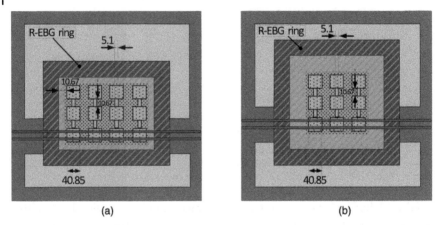

(a)   (b)

Figure 6.14 Top view of the removable EBG with (a) four vertical columns and (b) three vertical columns.

been developed, although a tuning process could be used to move the ring notch to a desired frequency.

6.3.3 Further Design Considerations

In addition to the parameters previously described for the design of a removable EBG-based CM filter, there are also other considerations to be taken into account in the project workflow. These additional considerations are related to all the issues that are not encountered in the standard onboard EBG filters design. The most important of these considerations are the impact of the metal ring around the EBG patches, the impact of the vertical vias used to create an electrical connection between the inner parts of the removable EBG and the main board, and finally some consideration about possible increase in the electromagnetic radiation from the EBG filter.

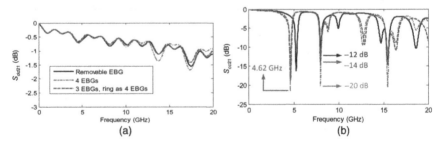

(a)   (b)

Figure 6.15 R-EBG filter with three and four columns of EBGs. (a) S_{dd21}. (b) S_{cc21} (Reproduced from Ref. [15] with permission of IEEE.)

6.3.3.1 Impact of the Metal Ring

The previous sections described the response of the removable EBG filter associated with the design parameters. It was observed that the presence of a metal ring laid out on the same layer as the EBG patches and surrounding the EBG cavities was able to maintain the required design frequency. This metal ring also generated some additional and, in general, beneficial (when considering the S_{cc21} response) further notches due to its resonance.

In this section, some insights into the global role of the ring will be given in depth. Therefore, two different and limit configurations, shown in Figure 6.16, are analyzed. The configuration in Figure 6.16a has the removable EBG without the presence of the ring, whereas the configuration shown in Figure 6.16b includes only the ring without the EBG. These are the two extreme opposite configurations. The $|S_{dd21}|$ and $|S_{cc21}|$ results for these configurations are compared in Figure 6.17 along with the case of the 'normal' removable EBG, that is, the complete removable EBG with the ring.

The "only-ring" case maintains the 5.26 GHz notch, but the lack of the EBG makes the 8 GHz notch disappear. Conversely, the case without the metal ring does not have the ring resonance, as expected, but at the same time it also reduces the EBG effectiveness. The notch depth decreases to −5 dB and it is shifted to 8.36 GHz.

Variations in ring width should also be considered and analyzed. For this study, the EBG geometry presented in Figure 6.9 is modified, making the metal ring larger: The ring is enlarged from 2g to 4g, as shown in Figure 6.18.

Figure 6.19 compares the S-parameter results for the two configurations with the ring width equal to 2g and 4g. The impact of the increased ring width is visible in $|S_{cc21}|$ in Figure 6.19b, where the first notch is shifted downward from

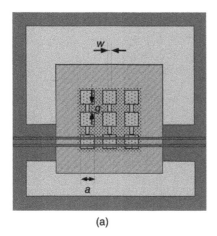
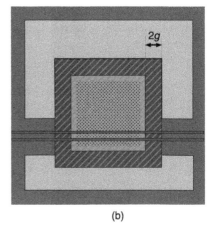

(a)   (b)

Figure 6.16 Top view of two different removable EBG-based CM filter cases. (a) R-EBG without metal ring. (b) Model with the ring (only ring) but without the EBG patches.

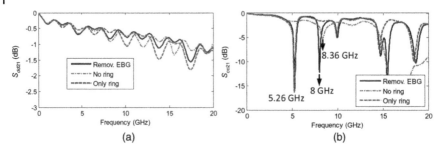

Figure 6.17 Impact of the metal ring on (a) $|S_{dd21}|$ and (b) $|S_{cc21}|$ of the removable EBG-based CM filter with dielectric constant $\epsilon_r = 14.4$. (Reproduced from Ref. [15] with permission of IEEE.)

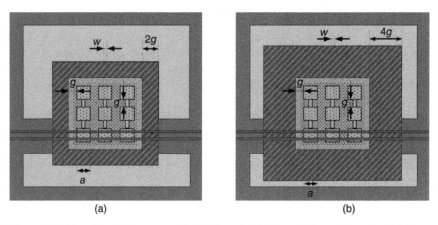

Figure 6.18 Top view of the removable EBG-based CM filter with two different metal ring width. (a) Width $= 2g$. (b) Width $= 4g$, g being the gap width.

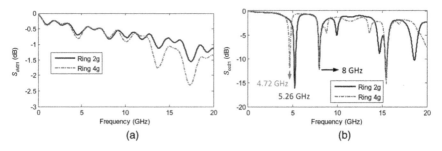

Figure 6.19 Removable EBG. Impact of the ring width on (a) $|S_{dd21}|$ and (b) $|S_{cc21}|$. (Reproduced from Ref. [15] with permission of IEEE.)

5.26 to 4.72 GHz. The presence of the metal ring should be taken into account in the design stage to compensate its downward shift of the resonant frequency. However, note that the filter notch frequency at 8 GHz is not affected by the ring width.

6.3.3.2 Impact of the Vias

Another important parameter in the design of a removable EBG-based CM filter can be found in the analysis of the effects induced by the vertical vias. These components are used in different ways, depending on the specific filter configuration. The two configurations where vias need to be considered are that (i) the data traces remain on the main board and only the EBG filter is on the removable part (the case considered thus far), or (ii) the data traces transition to the removable part and are in closer proximity to the EBG filter. For the first configuration where the data traces remain on the main board, the vias connect the ground-reference on the main board with the metal ring around the EBG patches, in order to establish a current return path on the removable board. Alternatively, in the R-EBG configuration with signal traces transitioning into the removable component, other vias connect the signal traces on the main board to the traces inside the removable filter itself. In the first configuration (in which the vias are connecting the main board reference layer with the metal ring only), a further configuration can be considered, that is, the configuration characterized by the absence of the metal ring. In this configuration, the return current path through the EBG must be considered, requiring the design to extend the vias (previously connecting the PCB with the ring) to reach the top metal layer of the R-EBG. This geometry is represented in Figure 6.20a.

The current path in Figure 6.20a is able to excite the EBG resonance. This assumption is validated by the results in Figure 6.21. As can be seen from Figure 6.21, the ring resonance at 5.26 GHz is missing when the ring is removed,

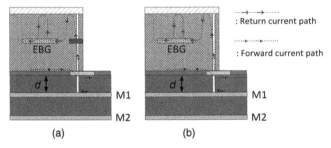

Figure 6.20 Removable EBG-based CM filter with and without metal ring considering an extended vias connection. (a) Original vias connection to the metal ring (only). (b) Extended vias without the metal ring.

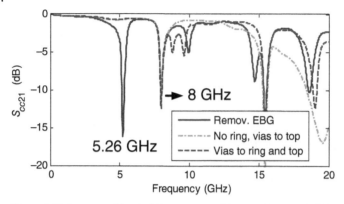

Figure 6.21 Impact on $|S_{cc21}|$ of the presence of the metal ring and of the extension of the vias to the top reference plane. (Reproduced from Ref. [15] with permission of IEEE.)

but also when the ring is connected to the upper metal layer (local ground-reference). Also, the desired notch at 8 GHz is unaffected by the missing ring, or when the ring is connected to the upper local ground-reference. When the ring is connected to the upper local ground-reference, the 5.26 GHz notch disappears since the ring is shortened at its four corners. The resonant frequency is shifted upward, thus generating additional notches slightly above the notch at 8 GHz.

6.4 Design Examples and Typical Results

The procedures and considerations previously introduced in this chapter are now applied to the design of more elaborate removable EBG CM filters. In particular, two specific design examples will be fully explained. The first consideration will address the reduction of the filter dimensions and therefore the space required on the main board. The second consideration will go into details concerning the trade-off between notch depth and notch bandwidth. This second consideration is usually one of the most important considerations during the EBG-based CM filter design process.

In order to provide a realistic scenario, the two examples will largely be based upon commonly used technologies and dielectric materials, such as an organic dielectric (specifically Megtron 6[1]) and a ceramic dielectric. The obtained results are very good and represent what is obtainable by a real-world EBG filter manufacturing.

---

1 http://www.matrixelectronics.com/products/panasonic/megtron-6/.

6.4.1 Design Example 1: Compactness Enhancement

As a first example, a removable EBG-based CM filter is developed and designed to save space on the PCB. The general design workflow previously described is maintained, but the space saving is achieved through the adoption of a new EBG geometry named "sandwich removable EBG," which was first introduced in Ref. [18].

Previous chapters have considered several different geometries of the EBG in order to reduce the EBG space consumption on the main board. This was achieved by manipulating the shape and dimensions of the patches. Among the proposed solutions, the 'sandwich geometry' seems to be very attractive to be implemented on a removable EBG-based CM filter. In practice, starting from the classical EBG solution based on arrays of patches in the onboard version, it is possible through the introduction of two additional layers and by eliminating the central patch (which is not essential [18]) to take one row of patches and to subsequently fold another row on a different layer in the PCB stack-up. In this process, the bridge connecting each pair of patches in the previous array structure is replaced by a vertical via that ensures the electrical connection between each of the resonant patches. The main result of this folding is the achievement of a smaller R-EBG size, since the occupied surface is related to only one single row of patches. The overall "sandwich" structure is depicted as shown in Figure 6.22a.

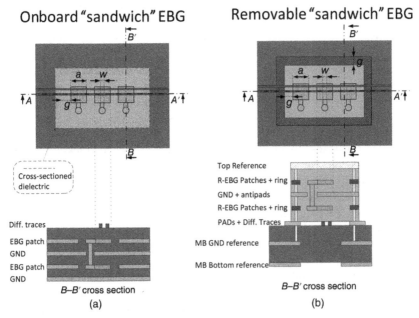

Figure 6.22 General geometry of an R-EBG with "sandwich" structure. Top view and stack-up of (a) onboard version and (b) removable version, respectively, with traces on the main board.

The sandwich configuration can be laid out either on the onboard configuration or on the removable EBG board, since the customization of the originally proposed sandwich geometry to a removable component is straightforward. There are no particular constraints to be observed for the design of the EBG cavity, except those already discussed. However, depending on the design requirements, there could be some variations with respect to the basic idea.

In particular, for the onboard implementation, it is convenient to place the EBG layers in such a way as to have one EBG layer directly below the signal differential traces and the other EBG layer between two reference layers, as shown in Figure 6.22a. However, in general, it is also possible to rearrange the layers in the most convenient way for the application of interest, as in the following example. The design process for a "sandwich" removable EBG-based CM filter, starting from its onboard counterpart, will result in a structure similar to the configuration in Figure 6.22b.

Again, the key point of any removable EBG-based CM filter allows the flexibility to alter the R-EBG stack-up without having to modify the main board. This flexibility is emphasized in the present design case, since the change of the PCB main board design would have a major impact on an onboard implementation; however, any change on an independent component would have no impact on the main board. Design Example 1 deals with the sandwich R-EBG solution, illustrated in Figure 6.23.

This design considers square EBG patches employed in a sandwich structure and customized to the particular case of traces entering inside the removable component and hence not remaining on the main board. The traces leave the PCB from pads and are then brought up vertically to minimize the discontinuity of impedance. Once the proper layer has been reached (in this example, the third from the top of the stack-up of the removable component), they travel as striplines until the descending vias are reached and the signals return to the main board. In this design, the dielectric material is Megtron6 and the target frequency is 8 GHz, similar to previous examples. Values for the stack-up dimensions, metal and dielectric thicknesses, are shown in Figure 6.23c.

A preliminary EBG design consists of EBG vias placed at the edge center of each square patch, as shown in Figure 6.22. However, the variation of the via position along the patch edge is able to alter the resonant frequency of each single EBG. Based on this consideration, a tuning process is applied moving the vias of the lateral EBGs toward the central EBG, thus still keeping a symmetric geometry, as shown in Figure 6.23b.

Electromagnetic simulations by using CST Studio Suite 2015 allows us to compute the common mode and differential mode insertion loss, shown in Figure 6.24. In this figure, the common mode response (S_{cc21}) shows the presence of a deep main notch at 8 GHz characterized by a quite large bandwidth (~675 MHz) computed at −10 dB.

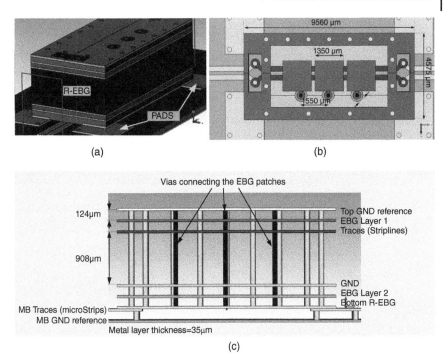

Figure 6.23 Structural details of the Design Example 1. (a) Perspective view of the entire removable filter. (b) Top view of the EBG layer 1. (c) Side view showing all the layers employed in this R-EBG solution. *Note:* In this case, the differential signal traces enter inside the removable EBG filter.

This is mainly due to the degree of asymmetry given by the vias placement that produces two resonant frequencies very close to each other at 7.9 and 8.18 GHz.

Moreover, the differential transmission coefficient (S_{dd21}) shows that the intentional differential signal is not negatively affected by the presence of the removable EBG-based CM filter. In order to better understand the cause of the various notches (above 8 GHz) found in the response of S_{cc21}, an analysis of the spatial distribution of the electric field at the S_{cc21} relevant frequencies is performed. The results are summarized in Figure 6.25. Two notch frequencies were observed above 8 GHz, one notch at 18.3 GHz and the another notch at 15.58 GHz. Two horizontal cut planes have been defined, corresponding to the locations of the largest magnitude of the electric field.

The first cut plane was set at between the main board and the bottom side of the removable filter since the small notch at 15.58 GHz seems due to an undesired resonance of the cavity formed by the removable component and the top reference plane of the main board. This resonance can become a

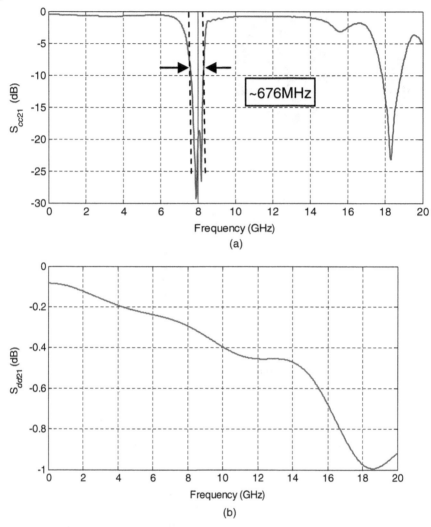

Figure 6.24 Frequency spectrum of the magnitude of (a) $|S_{cc21}|$ and (b) $|S_{dd21}|$ for the "sandwich" R-EBG CM filter in Figure 6.23.

detrimental source of radiation. The second cut plane was set between the first EBG layer and the underlying reference, since for the notch at 18 GHz, most of the electric field is located in this area. It appears that the cavity formed by the ring around the EBG layer 1 and the reference layer is able to induce a strong resonance for higher frequencies.

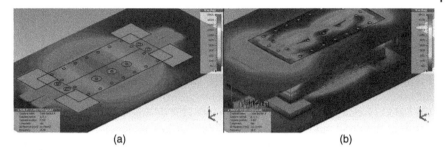

(a)   (b)

Figure 6.25 Spatial distribution of the electric field for different resonant frequencies.
(a) $f = 15.58$ GHz, cut plane between the main board and the bottom of the removable filter.
(b) $f = 18.3$ GHz, cut plane between EBG layer 1 and the R-EBG stripline layer.

Figure 6.26 shows the total radiated power from the R-EBG configuration. A comparison is made between the case where there are only the microstrip traces on the main board (no EBG) and the case in which the sandwich removable EBG-based CM filter is used. An increase of the total radiation is evident from 6 GHz on with an average difference of approximately 20 dB. This additional radiated energy must be contained by the overall shielding of the system.

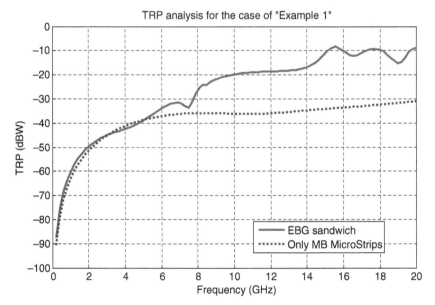

Figure 6.26 Total radiated power for the design Example 1 (EBG sandwich) compared with the baseline with only the microstrips on the main board.

6.4.2 Design Example 2: Filtering Features Enhancement

A second design example of a removable EBG-based CM filter is discussed with the aim to obtain an optimal trade-off between notch depth at the design frequency and the filter notch bandwidth. The design procedure is fundamentally the same discussed previously; therefore, for this example, we will focus on the aspects strictly related to the attainment of the desired trade-off between the depth of the notch and the bandwidth. Without loss of generality, the R-EBG-based filter will be developed on an LTCC substrate to show the flexibility and generality of the proposed design approach.

In this second design example, the major effort will address the proper sizing of the EBG resonant cavities in order to achieve a filtering notch as wide and at the same time as deep as possible. Some further issues observed in the previous design example are also addressed, for example, the requirement that the filter should be physically small and that unwanted resonances (possible candidate sources of electromagnetic radiation) should be eliminated or reduced as much as possible.

For this example, the filter configuration will employ a set of solid EBG rectangular patches without bridges. This choice is motivated by the fact that a larger number of narrower patches with slightly different lengths generate cavities resonating at different frequencies, which are close to each other. Therefore, the S_{cc21} response would feature an overall larger common mode filtering effect. In Ref. [5], the miniaturization constraint led to introduce the EBG type of patch from the larger solid cavity; herein the trend goes back to the solid patch of rectangular shape, with its long dimension to be designed according to the target resonant frequency, and its short dimension to allow several patches to fit within the original area dedicated to the EBG filter, thus without affecting the overall filter size constraints.

The basic structure for the proposed filter is depicted in Figure 6.27. The EBG top view accounts for a generic number of rectangular patches arranged in an array. The specific number of patches will be chosen according to the requirements of the application, in terms of the manufacturing technology (employed stack-up, minimum separation, building process, etc.), the maximum surface available on the main board for the EBG-based filter surface, and so on. The chosen configuration for the filter under analysis requires that the differential signal traces enter inside the removable EBG and run as striplines within the removable component.

Note in Figure 6.27 the presence of two EBG layers directly above and below the internal stripline traces. This design choice should allow us to achieve a deeper filter notch.

The relevant geometrical details of this removable EBG-based CM filter are shown in Figure 6.28. The configurations in Figure 6.28a and b show two different solutions for the reference pads design on the main board. Figure 6.28a

R-EBG structural top view

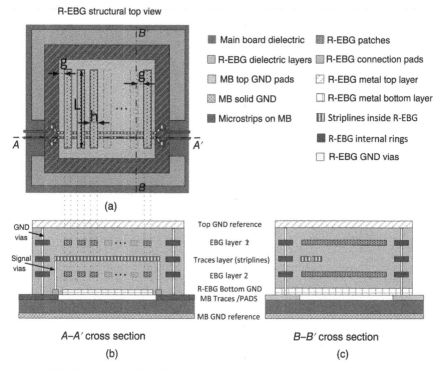

Main board dielectric   R-EBG patches

R-EBG dielectric layers   R-EBG connection pads

MB top GND pads   R-EBG metal top layer

MB solid GND   R-EBG metal bottom layer

Microstrips on MB   Striplines inside R-EBG

R-EBG internal rings

R-EBG GND vias

(a)

GND vias   Top GND reference

   EBG layer 1

Signal vias   Traces layer (striplines)

   EBG layer 2

R-EBG Bottom GND
MB Traces /PADS

MB GND reference

A–A′ cross section   *B–B′* cross section

(b)   (c)

Figure 6.27 Structural details of the design Example 2 developed for notch depth–width enhancements. (a) Top view of the EBG layer 1. (b and c) The side views of the stack-up.

shows four reference pads placed at the four corners of the removable component. Figure 6.28b shows two large patches connecting the vias coming from the same side of the removable component. The choice of these reference pads design will affect some of the final design results.

The solution with four reference pads on the corners of the R-EBG is affected by an unwanted resonance occurring between the R-EBG and the main board (as seen in the first design example). Alternatively, the solution with two large reference pads on either side of the removable component eliminates this unwanted resonance, thereby reducing the radiated power. Furthermore, as can be seen from Figure 6.28c, the model is characterized by several staggered vias to connect the metal planes. Staggered vias are typically required owing to their mechanical properties of a design based on LTCC substrates. The number of resonant rectangular EBG patches is equal to five.

In order to show the main features of this type of R-EBG filter, two different options are analyzed. The first option is without any notch enlargement features, and the second option includes features to increase the filter notch bandwidth.

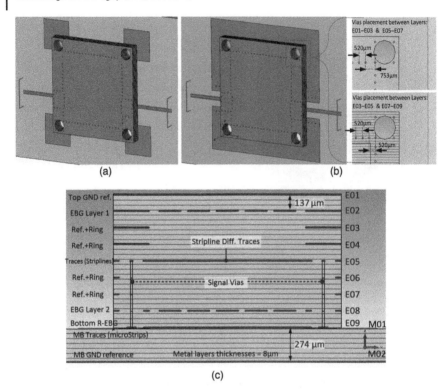

Figure 6.28 Details of pads design. (a) Perspective view of a solution with four reference pads at the corners. (b) Perspective view of a solution with two large surrounding pads. (c) Side view of the stack-up.

The basic design step for both filter configurations consists of determining the geometrical dimensions of the EBG patches. The length of the patches is determined according to the desired filter notch frequency, following the expression in Equation (4.2) defined for square solid patch, herein to be applied to the long rectangular patch dimension. Other considerations to be taken into account are related to issues such as the adjustment of the differential impedance of the signal traces in order to reduce the impedance discontinuity when the signals enter the R-EBG. This is done through the optimization of the stripline traces cross-sectional dimensions and the proper sizing of the vertical vias transitions; in particular, a TDR-based simulation is employed for evaluating the differential impedance of the trace path across the transition through the removable filter section. A few step trial-and-error tuning of the via and stripline spacing allows minimizing the discontinuities of the differential impedance. Another parameter that can be set is the width of the rectangular solid patches: This is selected in the example provided herein to fit five rectangular cavities within the original EBG area, a number larger than the three EBGs originally included in the removable filter shown in Figure 4.18.

For this particular example, the dielectric is an LTCC material, having a relative dielectric constant $\epsilon_r = 7.85$ and a target differential impedance for the signal traces equal to $100\,\Omega$.

We will again use the same target frequency of the preceding case, that is, 8 GHz for the removable EBG-based CM filter. Using the equations developed in Chapter 4, the R-EBG dimensions will be as follows. The microstrips on the main board are $170\,\mu m$ wide with a separation distance of $150\,\mu m$, and the striplines running inside the R-EBG are $150\,\mu m$ wide with a separation of $450\,\mu m$. The width of the rectangular EBG patches is set to $h = 1400\,\mu m$; and their length is $L = 6250\,\mu m$, as shown in Figure 6.29a. For the standard case, all the patches are equal; in the larger bandwidth (LBW) solution, instead, the central patch has the same length of $L3 = 6250\,\mu m$, whereas the other patches have been manually tuned in order to split the main notch into a wider set of frequencies: The longest EBG patch (of length $L1$) is $6615\,\mu m$ long and the shortest ($L4$) is $6075\,\mu m$ long, as shown in Figure 6.29b. The other patch lengths are $L2 = 6350\,\mu m$, and $L5 = 6125\,\mu m$. The manual tuning process provided as the best response the configuration where $L4$ is slightly smaller than $L5$.

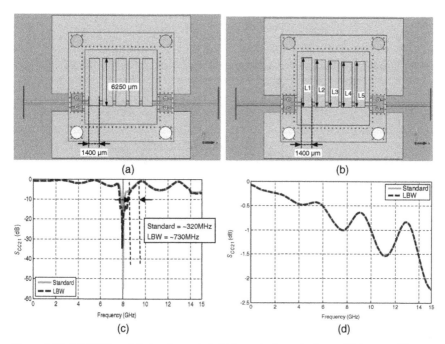

Figure 6.29 R-EBG filter of design Example 2. (a) Layout of standard case. (b) Layout of larger bandwidth (LBW) case. (c and d) $|S_{cc21}|$ and $|S_{dd21}|$, respectively, for both cases.

The results of the magnitude of S_{cc21} and S_{dd21} are shown in Figure 6.29c and d. As can be seen from the obtained simulation results in Figure 6.29c, the filter notch results for the larger bandwidth version more than double with respect to the standard case. The evaluation of the bandwidth is done at a reference value of −10 dB. In both designs, the differential mode is not impacted by the presence of the wider bandwidth component.

From the point of view of the total radiated power (TRP), the developed filters exhibit the behavior shown in Figure 6.30. The TRP is plotted for different configurations as described in the caption, placing together the TRP responses of both the Example 1 cases and the Example 2 model responses. The results of Example 1, already shown in Figure 6.26, show an increase of radiation of more than 10 dB starting from the filtering frequency. In the Example 2 comparisons, instead, the radiation increase is smaller with few frequency bands in which the difference reaches almost 10 dB. Example 2 provides a better shielding with respect to Example 1, due to a denser stitching via fence and blind vias without antipads on outer layers (as occurring in Example 1, as shown in Figure 6.23a). However, in both examples, the TRP level increases when the removable filter is mounted on the main board. This drawback will be investigated in Chapter 7 where reverberation chamber measurements help identifying the appropriate shielding configuration for making the EBG filter effective from common mode current reduction (S_{cc21}) as well as from a radiated EMI viewpoint.

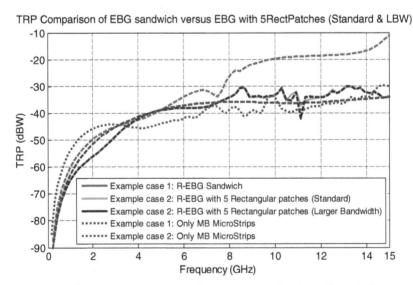

Figure 6.30 Comparison of the total radiated power for the design Example 1 (sandwich R-EBG) and for the design Example 2 (standard case and larger bandwidth case) and for the case without the removable EBG on the main board (only MB).

6.5 Summary of Advantages and Drawbacks

The removable EBG concept is powerful and can help many potential EMC issues. However, it is important to highlight the main advantages and drawbacks concerning the use of R-EBG CM filters. The advantages of using R-EBG filters are in (a) layout flexibility and PCB scalability compared to the onboard implementation, (b) performance obtained in terms of common mode and differential mode insertion loss, and (c) total radiated power.

Advantages from the Viewpoint of the Layout Flexibility

- The removable EBG-based CM filter can be installed, substituted, and removed by the typical soldering assembly process with ease in order to meet the requirements of PCB differential link manufacturers during any stage of the design process. The device can be easily scaled to different target frequencies by simply modifying the EBG geometry or the dielectric material.
- The R-EBG target frequency can be easily modified without the need to modify the main board. The standard onboard EBG-based CM filters cannot be easily scaled to different target frequencies since the main board must be modified often with significant expense.

Advantages from the Viewpoint of the Filtering Performance

- The removable filter presents more degrees of freedom for the enhancement of the main notch, in terms of both depth and bandwidth, since the possibility to easily modify the removable stack-up allows the design of more elaborated EBG patch structures without huge costs.
- As highlighted from the sandwich R-EBG of design Example 1, it is possible to obtain good filtering performance together with required surface reduction on the main board.

Main Drawbacks

- The vertical via transitions used to electrically connect the removable part to the signal traces are source of differential impedance variation and hence possible degradation of S_{dd21}. Therefore, the design of the R-EBG requires more effort to reduce this effect than the onboard implementation of an EBG.
- The inherent additional geometries that are part of the R-EBG filter can excite further resonances in the structure (as seen in the case of the external ring in the design Example 1). This can create more than the desired notch in the S_{cc21} behavior potentially at the cost of introducing unwanted radiation.
- The total radiated power at the resonant frequencies is clearly high. This problem can be addressed in different ways. Some approaches will be

discussed in Chapter 7, including absorbing materials able to mitigate unwanted radiation from the filter as well as radiation containment by a proper board shielding.

References

1 T.L. Wu, J. Fan, F. de Paulis, C.D. Wang, A. Ciccomancini, and A. Orlandi, Mitigation of noise coupling in multilayer high-speed PCB: state of the art modeling methodology and EBG technology. *IEICE Trans. Commun.*, Vol. E93-B, No. 7, 2010, pp. 1678–1689.

2 S. Shahparnia and O.M. Ramahi, A simple and effective model for electromagnetic bandgap structures embedded in printed circuit boards. *IEEE Microw. Wirel. Compon. Lett.*, Vol. 15, No. 10, 2005, pp. 621–623.

3 A. Ciccomancini Scogna and A. Orlandi, Multi band noise mitigation in PWR/GND plane pairs by using a 2D EBG structure with square patches and meander lines, in *Proc. of the 20th International Zurich Symposium on EMC, Zurich, Switzerland*, January 12–16, 2009.

4 F. de Paulis, A. Orlandi, L. Raimondo, B. Archambeault, and S. Connor, Common mode filtering performances of planar EBG structures, in *Proc. of the IEEE International Symposium on Electromagnetic Compatibility*, 17–21 August, 2009, pp. 86–90.

5 F. de Paulis, B. Archambeault, M.H. Nisanci, S. Connor, and A. Orlandi, Miniaturization of common mode filter based on EBG patch resonance, in *Proc. of the IEC DesignCon 2012*, January 30–February 2, 2012, Santa Clara, CA.

6 F. de Paulis, L. Raimondo, S. Connor, B. Archambeault, and A. Orlandi, Compact configuration for common mode filter design based on electromagnetic band-gap structures. *IEEE Trans. Electromagn. Compat.*, Vol. 54, No. 3, 2012, pp. 646–654.

7 M.H. Nisanci, F. de Paulis, A. Orlandi, B. Archambeault, and S. Connor, Optimum geometrical parameters for the EBG-based common mode filter design, in *Proc. of the 2012 IEEE Symposium on Electromagnetic Compatibility, Pittsburgh, PA*, August 5–10, 2012.

8 T.-L. Wu, H.-H. Chuang, and T.-K. Wan, Overview of power integrity solutions on package and PCB: decoupling and EBG isolation. *IEEE Trans. Electromagn. Compat.*, Vol. 52, 2010, pp. 346–356.

9 L. Raimondo, F. de Paulis, and A. Orlandi, A simple and efficient design procedure for planar electromagnetic bandgap structures on printed circuit boards. *IEEE Trans. Electromagn. Compat.*, Vol. 53, No. 2, 2011, pp. 482–490.

10 F. de Paulis and A. Orlandi, Accurate and efficient analysis of planar electromagnetic band-gap structures for power bus noise mitigation in the GHz band. *Prog. Electromagn. Res. B*, Vol. 37, 2012, pp. 59–80.

11 F. De Paulis, B. Archambeault, S. Connor, and A. Orlandi, Electromagnetic band gap structure for common mode filtering of high speed differential signals, in *Proc. of the DesigCon 2011, Santa Clara, CA*, January 31–February 3, 2011.

12 A. Ciccomancini Scogna, A. Orlandi, and V. Ricchiuti, Signal and power integrity analysis of single ended and differential lines in multilayer PCBs with embedded EBG structures, on *IEEE Trans. Electromagn. Compat.*, Vol. 52, No. 2, 2010, P 357-L 364.

13 X. Gu, R. Rimolo-Donadio, Y.H. Kwark, C. Baks, F. de Paulis, M.H. Nisanci, A. Orlandi, B. Archambeault, and S. Connor, Design and experimental validation of compact common mode filter based on EBG technology, in *Proc. of the IEC DesignCon 2013*, January 28–31, 2013, Santa Clara, CA.

14 F. De Paulis, L. Raimondo, D. Di Febo, B. Archambeault, S. Connor, and A. Orlandi, Experimental validation of common-mode filtering performances of planar electromagnetic band-gap structures, in *Proc. of the 2010 IEEE Symposium on Electromagnetic Compatibility, Fort Lauderdale, FL*, July 25–30, 2010.

15 F. de Paulis, M.H. Nisanci, D. Di Febo, A. Orlandi, S. Connor, M. Cracraft, and B. Archambeault, Standalone removable EBG-based common mode filter for high speed differential signaling, in *Proc. of the IEEE International Symposium on EMC, Raleigh, NC*, August 4–8, 2014.

16 M.A. Varner, F. de Paulis, A. Orlandi, S. Connor, M. Cracraft, B. Archambeault, H. Nisanci, and D. di Febo, Removable EBG-based common mode filter for high speed signaling: design and experimental validation, in *IEEE Trans. Electromagn. Compat.*, 2015.

17 C.-Y. Hsiao, C.-H. Cheng, and T.-L. Wu, A new broadband common-mode noise absorption circuit for high-speed differential digital systems. *IEEE Trans. Microw. Theory Tech.*, Vol. 63, No. 6, 2015, pp. 1894–1901.

18 C. Olivieri, F. de Paulis, A. Orlandi, S. Connor, and B. Archambeault, Miniaturization approach for EBG-based common mode filter and interference analysis, in *Proc. of the IEEE International Symposium on EMC, Santa Clara, CA*, March 15–20, 2014.

19 PCI Express Base 3.1 Specification. Available at http://www.pcisig.com/specifications/pciexpress/base3/.

20 Computer Simulation Technology, CST Studio Suite 2015. Available at www.cst.com.

21 F. de Paulis, M.H. Nisanci, and A. Orlandi, Experimental validation of an 8 GHz EBG based common mode filter and impact of manufacturing uncertainties, in *Proc. of the IEEE Symposium on Electromagnetic Compatibility, Denver, CO*, August 5–9, 2013.

7

EBG Common Mode Filters: Modeling and Measurements

with contribution of Michael Cracraft

In this chapter, we will explore the practical considerations of designing and executing experiments to validate the operation of the EBG designs presented in the previous chapters. A successful common mode filter design will have little or no impact on the intentional differential signal, will suppress common mode signals by a designed amount over a defined bandwidth, will not increase cross talk to a detrimental level, and will not increase radiated emissions such that the benefits of reducing the common mode signal is negated. In order to validate an EBG design, one must design a test vehicle that enables all of these metrics to be measured reliably.

The most straightforward approach for measuring how well an EBG filter passes differential and common mode signals is to use a four-port vector network analyzer (VNA) to measure the scattering parameter (S-parameter) matrix. Other approaches do exist, such as using time domain reflectometry/time domain transmission (TDR/TDT) to characterize the time domain response of the filter and then mathematically converting this response to an S-parameter representation, and you can choose the approach that suits the equipment and analysis tools that you have. We will focus on VNA measurement approaches in this chapter, because this is our preferred technique.

Several measurement environments exist for measuring radiated electromagnetic interference (EMI) emissions from the filter, including far-field measurements in open area test sites or semianechoic chambers, near-field measurements with scanning rigs or hand-held probes, and radiated power measurements in gigahertz transverse electromagnetic (GTEM) cells and reverberation chambers. Each of these measurement approaches has a set of advantages and disadvantages, and we will spend some time discussing these in order to provide context for making decisions on future real-world cases. The measurement techniques themselves are well defined in various test standards, so not much will be said about those details. The more interesting and

Electromagnetic Bandgap (EBG) Structures: Common Mode Filters for High-Speed Digital Systems,
First Edition. Antonio Orlandi, Bruce Archambeault, Francesco De Paulis, and Samuel Connor.
© 2017 by The Institute of Electrical and Electronics Engineers, Inc. Published 2017 by John Wiley & Sons, Inc.

challenging aspect of making these measurements is determining how to design the test structure such that the emissions from the filter are distinct from other sources.

As we discuss the design and measurement of various EBG filters in this chapter, we will expand on the concepts above and try to provide enough technical details so that you can make good decisions in your own work.

7.1 Design Considerations for the EBG Filter Test Fixture

In this section, we will discuss the considerations associated with designing the test fixture used to measure EBG filter designs. The first issue is deciding where the boundary between the test fixture and the device-under-test (DUT) is going to be. Next, we will study options for the launch structure. De-embedding options will be discussed throughout, so that it becomes clear how intertwined the design and de-embedding technique are. Finally, we will talk briefly about attachment/assembly considerations and how these choices can affect measurement repeatability and test efficiency.

7.1.1 Defining the Bounds of the EBG Device

At first it might seem like a trivial question, but in fact there are important subtleties to consider when answering the question: Where does my EBG filter device start and end? Regardless of whether the filter is embedded inside a PCB or mounted on top as a surface-mount part, it is important to define the location of the interface between test fixture and device. This is critical because we want to subtract any effects of the launching structure in order to get an S-parameter matrix that corresponds to the filter device only. This subtraction process is known as de-embedding, and there are several techniques published in the literature and encoded into various commercial tools that can effectively remove a fixture's response. Each of the de-embedding techniques makes assumptions about the interface location, and therefore the test fixture must be designed with a specific de-embedding technique in mind.

For most of the sample designs we have explored, we opted for an interface where the differential pair has a well-defined transmission line mode and impedance established, which means being a sufficient distance away from the discontinuities of via transitions or the EBG structure itself. The trade-off is added insertion loss for higher measurement fidelity. The final filter response will include some length of differential transmission line before and after the EBG structure, but the chosen interface plane will support very high-quality de-embedding, which means higher measurement repeatability and less measurement uncertainty. This trade-off is well-made if you can validly assume that the

Figure 7.1 CAD model of a removable EBG device mounted on a test board with the location of a de-embedding plane shown.

insertion loss of the extra trace length is negligible. The placement of the interface plane can be quite challenging in some cases, for example, when characterizing high-speed connectors, because it is extremely difficult to say where the connector stops and the PCB begins. The true interface is somewhere inside a plated via or through-hole, but the environment is so electromagnetically complex that basic de-embedding assumptions are invalid. Figure 7.1 shows where the interface plane might be placed for a removable EBG device. The interface plane is indicated by a translucent "window" that is placed a short distance before the differential pair passes underneath the EBG device and connects to the surface-mount pads. This location has the de-embedding advantages described above, but also has simulation advantages. One can place a wave port in the model at the illustrated location, because the transmission line modes are well defined and there is a consistent two-dimensional cross section across the model. The simulation results will only include signal transmission effects in the structure between the wave ports, which means the simulation results should match well with de-embedded measurement results [1].

7.1.2 Selecting a Launch Structure for the Test Fixture

Once the bounds of the EBG device are determined, the remaining fixture design problem is how to transition from the 50 Ω, single-ended, coaxial environment (used by nearly all measurement equipment) to the typically non-50 Ω, differential, stripline (or microstrip) environment that exists at the interface plane of the EBG device. This transition region is often called the "launch structure," because it takes the signals from the coaxial cables and launches them into the DUT. While easy to conceptualize, it is very difficult in practice to design this launch structure well for high-frequency PCB applications.

There are several approaches that a test fixture designer might take, but in all cases the designer must try to minimize impedance discontinuities to avoid large reflections and excessive attenuation. Reflections and attenuation are detrimental, because they both reduce the amount of energy that is launched into the EBG, which reduces the dynamic range of the measurement. In fact, problems in the launch structure have a doubled impact, because any reflections from the DUT itself (which we are interested in seeing) have to transmit back through the launch structure to get to a measurement port. This reflected signal then experiences the same attenuation and scattering that the original launched signal saw. Some of the potential sources of impedance mismatch in the launch structure are the PCB coaxial connectors, the vias and antipads in the PCB, and the transmission line cross section in the PCB. We will discuss each in more detail.

PCB coaxial connectors come in various types and grades, and the bandwidth and manufacturing tolerances vary accordingly. For measurements up to 40 GHz, a 2.92 mm compression-fit coaxial connector is a popular choice. This connector has a reference footprint you can place on a board design, and this footprint will keep the impedance as close to 50 Ω as possible throughout the transition from the connector to the PCB surface pads. If you are designing for higher or lower frequencies, there are corresponding coaxial connectors that will provide the desired performance (see Table 7.1). There is no point in using more expensive components than you need, but you must also be aware of factors that will limit the maximum usable frequency despite the rating of the

Table 7.1 List of coaxial connectors in order of performance and cost (highest to lowest).

Connector/ connection type	Maximum frequency (approximate)	Comments
1.0 mm	110 GHz	Highest-performance precision coaxial connector; not compatible with other types
1.85 mm	65 GHz	Compatible with 2.4 mm connectors; precision connector available in instrument and metrology grades
2.4 mm	50 GHz	Compatible with 1.85 mm connectors; precision connector available in instrument and metrology grades
2.92 mm	40 GHz	Compatible with 3.5 mm and SMA; precision connector available in instrument and metrology grades
3.5 mm	26.5 GHz	Compatible with SMA and 2.92 mm connectors; dimensions are more tightly controlled than SMA
SMA	18–26.5 GHz	Widely used; inexpensive; quality varies; offered in many configurations (through-hole, compression-fit, right-angle, edge-mount)

connector family. For example, through-hole SMA connectors will typically be a poor impedance match to your PCB due to the size and spacing of the four "ground-reference" leads. Some right-angle SMA connectors have a very high VSWR due to the bend in the coaxial structure. Variation in connectors due to mechanical tolerances will introduce errors in de-embedding, because your reference structures will not match your experiment structures. For the most recent work on filters for 8 and 10 GHz, the authors have used 2.92 mm connectors and had very good results.

After selecting the coaxial connectors, the next challenge is determining how to wire from the surface pad to the DUT. If you opt to route your test fixture with microstrip wires, you can avoid via stub effects by mounting the coaxial connector on one side of the PCB and running a via from the center conductor's pad to the microstrip trace on the opposite side of the board. Designing this transition via so that it matches 50 Ω is, however, not a simple task. The via barrel size, the pad and antipad sizes, the number of return vias, and the size and distance of the return vias will all have an effect on the impedance of the transition via. A designer should spend a significant amount of time simulating the connector footprint and transition via configuration and studying geometrical sensitivities before building a test fixture.

If we assume that the coaxial connector is well behaved and that the footprint and the transition vias maintain a 50 Ω impedance through the connector-to-PCB transition, then we still must wire the card in such a way that we can transition from 50 Ω single-ended traces to 85 Ω (or another target impedance) differential traces. We do not want to run 50 Ω single-ended wires all the way up to the EBG interface, because the de-embedding techniques assume and require that well-defined transmission line modes have been established at the interface plane. If we route the wires as microstrips, we can avoid via stub effects (and possibly avoid vias altogether), and microstrips have less dielectric loss per unit length. On the other hand, routing the wires as striplines will typically achieve better impedance uniformity and isolation from ambient noise. In the end, the choice of routing is usually determined by the EBG structure and how it is intended to be used in a real circuit.

Let us discuss a few realistic examples and the reasons why we would pick microstrip or stripline routing for those cases [2]. First, if we are trying to validate one of the removable EBG part designs, we would typically use microstrip wiring from the launch structure to the EBG pads because the differential pair has to be on the surface layer to connect with the EBG pads [1,3]. In making this choice, we also get the benefits of avoiding unnecessary via transitions, lower attenuation per unit length, and enabling the use of edge connectors or microprobes. Edge connectors are a type of coaxial connector that can be mounted to the edge of a circuit board, with the center conductor soldered to the microstrip trace directly and the outer conductor leads soldered to ground-reference pads. Microprobes can achieve

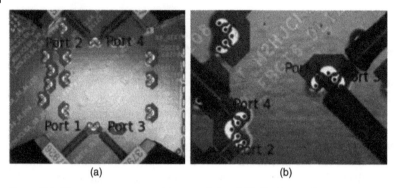

(a)   (b)

Figure 7.2 Measurement setup with ground-signal (G–S) microprobes connecting to a G–S–G–S–G land pattern. (a) Top view, showing how four ports can be probed. (b) Close-up view of the G–S–G–S–G land pattern with microprobes in place.

a very high-quality launch directly onto the microstrip traces, but they are expensive and fragile and require significant space and right-of-way to place them (see Figure 7.2).

As a second example, let us assume we are validating an EBG filter that is designed to be placed on high-speed differential pairs that are routed as striplines between the driver and the receiver or connector. In this case, a stripline launch is more appropriate, because it better mimics the real-world structure and will have the advantages of better impedance control and noise isolation. When using stripline routing, care must be taken with via stubs, and back-drilling might be required if stubs are longer than 1/10 wavelength at the highest frequency of interest. One way to avoid back-drilling is to mount the coaxial connectors on the top layer and run the striplines on layer $N-1$ with solid reference planes on layers $N-2$ and N, where N is the number of layers in the test board. This will keep the via stub as short as possible (around 0.125 mm for common stack-up heights, which is approximately as short as a stub could be with back-drilling).

7.1.3 Calibration Versus De-embedding

Now that we have discussed various physical details of the launch structure design, we should discuss options for removing the launch structure effects from a measured (or simulated) response. One's first instinct might be to use a calibration technique such as short, open, load, through (SOLT) or through, reflect, line (TRL) to remove the effects of everything up to the interface plane where the DUT begins. If possible, this would produce a very high-quality, reproducible measurement. The key idea in these calibration techniques is that if you place a known termination condition on your measurement ports, then you know what the ideal response should be. For example, the "through cases" should have no attenuation and no phase progression because your two measurement ports are connected directly to each other. Any deviation from this ideal response is "error" in the cables and connectors and can be mathematically removed with error vectors that are stored as part of the calibration. As long as you continue to use the same cables and connectors and their characteristics do not drift due to changes in temperature, humidity, positioning, or other factors, the error correction in the calibration will be effective at subtracting out their effects and measuring the response of the DUT only. If you want to use one of these calibration techniques for a PCB measurement, the main challenge is creating well-known termination conditions. For coaxial calibrations, there are off-the-shelf calibration kits with broadband shorts, opens, and $50\,\Omega$ loads that have been characterized up to some specified frequency. If you design your own calibration standards in your test board, then *you* must determine their true frequency-dependent impedances, either by measurement or by simulation. Any manufacturing tolerance issues or variations in dielectric properties will introduce error into the calibration, which will also propagate to your measurements. TRL calibration does not require a broadband load, but instead requires transmission lines of multiple lengths. The success of this technique is highly dependent on uniformity between the various transmission lines used for TRL and the transmission lines for the measurement cases themselves. Small differences in linewidths and dielectric properties (permittivity, thickness, fiber density, etc.) will introduce errors in the calibration that make measurement data very poor. When the authors have tried to use TRL calibration on lower cost PCBs with standard grade SMA connectors, the results were never acceptable above $10\,\mathrm{GHz}$.

Given the challenges associated with creating your own calibration sites (for either SOLT or TRL) on lower cost test boards, most designers opt for using an off-the-shelf SOLT calibration kit and calibrating at the ends of a set of good coaxial cables. This decision requires another approach to remove the effects of the launch structure. The term used across the industry for this is "de-embedding," because the launch effects are embedded in the measured s-parameters and need to be de-embedded to get the response of the DUT by itself. Conceptually, de-embedding is done by measuring the s-parameters of the launch by itself so that its response can be mathematically removed from the overall response, which includes launch plus DUT plus launch. A mathematical description of this de-embedding process is

$$\left[T_{launch}^{-1}\right]\left[T_{overall}\right]\left[T_{launch}^{-1}\right] \tag{7.1}$$

where the T_{overall} block is the scattering transfer parameters (T-parameter) representation of the overall structure (launches plus DUT) and the T_{launch}^{-1}

block is the inverse of the T-parameter matrix for the launch structure. Equation 7.1 shows the T^{-1}_{launch} block as identical on the left- and right-hand sides. In reality, the launch structures might not be identical for all ports and care must be taken when hooking up the ports of the T-parameter blocks (think of the difference between right-inverses and left-inverses in linear algebra). A similar mathematical approach can be taken with chain parameters (also known as $ABCD$ parameters).

There are quite a few different de-embedding approaches (structures plus algorithms) described in the literature, and we will only discuss a few examples here. The corresponding reference structures fall into two basic categories: 1×-reflect and 2×-through. The 1×-reflect is a launch structure that terminates into an open or short circuit at the location where the DUT interface plane would be. If the left- and right-hand side launches are different, you need separate 1×-reflects for each side. The 2×-through is a combination of the left- and right-hand side launch structures with no DUT in between. A popular commercial de-embedding solution is Automatic Fixture Removal© (AFR) by Keysight [4]. This algorithm will take data from a 1×-reflect, a pair of 1×-reflects for left and right sides, or a 2×-through measurement and de-embed the launch effects from measurement data of the full structure. Smart fixture de-embedding [5] is another de-embedding algorithm. SFD uses a 2×-through measurement and a waveform peeling algorithm that retains information about the differential impedance at the middle of the structure. It should be clear now that regardless of your choice of calibration or de-embedding to remove the launch structure effects, you will need to place special structures on your test board. In the case of AFR or SFD this would be a single 2×-through structure for every wiring layer of interest, and the 2×-through should be placed somewhere close to the structures of interest to increase the likelihood that the dielectric thickness and etching conditions are equivalent.

7.1.4 Physical Design Considerations

After the launch structure has been designed, the connectors have been selected, and the routing layers established, a designer is left with the final task of putting all of these together into a physical design that can be built and measured. There are limitations on how thin the test board can be, for example (somewhere between 1 and 1.5 mm is typical), and this can impact the dielectric thickness you can have in your routing layers. Another consideration is how close together you can place the launches for a single test site and how close together you can place independent test sites (a test site is an area of a test board on which you locate four launch structures and one EBG filter design permutation). For cost reasons, it is desirable to keep the test sites as small as possible and to pack them as closely together as possible. This allows one to get more design permutations into a single PCB panel. PCB panels are typically

$45.7 \times 61.0 \, \text{cm}^2$ with a usable area of $40.6 \times 55.9 \, \text{cm}^2$. The competing constraints are logistical: If you place the launch structures too close together, you will not be able to attach coaxial cables to the connectors. Depending on the size of the cable jackets and the strain relief mechanism, most cables used require a center-to-center spacing of $1-1.5$ cm between adjacent coaxial connectors. This allows enough space for the cables to be inserted vertically (without lateral forces that can damage the center conductors) and to be tightened with a torque wrench. For designs that use a microprobe launch structure, you can get two probes into a very small area (about $1 \, \text{mm}^2$ for a G–S–G–S–G launch), so tighter packing is possible.

While the width of a test site can be shrunk by using a microprobe, the length of the test site can never be shorter than the 2×-through length plus the length of the EBG filter structure, and the 2×-through length is limited by the de-embedding technique chosen. For example, SFD requires the 2×-through to be long enough so that the differential impedance settles to its true value in the middle of its TDR response. This length will depend on the loss in the transmission line, and the differential impedance difference between the single-ended and differential regions of the 2×-through. Simulating your design before building it is the best way to ensure that the length is sufficient. Figure 7.3 shows a 2×-through structure that worked well. The 2.92 mm connectors were separated by 1 cm in width and 2 cm in length. This provided a 1.5 cm length for the differentially routed section, which was just long enough for SFD to work (meaning the impedance in the TDR response settled down from near $100 \, \Omega$ in the single-ended regions to $85 \, \Omega$ in the center of the differential region).

Another logistical consideration is how the test site will be stabilized during measurements. If the test site is not secured in a fixed position, the coaxial cables

Figure 7.3 2×-through structure from a test card design. Note the complex via arrangement for each 2.92 mm connector, the 45 degree turns on the routing, and the rapid transition from single-ended into differential routing. These design choices produced a response that worked well with SFD de-embedding.

will be free to flex and move. While some degree of movement should not affect your measurements (if you do see fluctuations in your measurement with cable movement, this is a sign that the connections are not torqued correctly or that the cable itself might have a problem), changing the position of the test card and/or the bends in the cables will hurt the repeatability of your measurements and introduce error into comparisons between test sites. For this reason, it is preferable to stabilize your test board in some way. Obvious choices are clamping it with a vise, screwing it to a stiffener plate or similar mechanical support, or holding it to an air table with a vacuum pump. Depending on your choice, you might need to provide extra area around your coaxial connectors for mounting holes or clamp keepouts. If your mounting or clamping hardware is conductive and is placed too close to the launch structures, there will be parasitic capacitance that can affect the impedance of the transition from the cable to the PCB. This effect can be quantified with simulations to determine at what distance it becomes negligible. Nonconductive clamping hardware and stiffener plates may be used without influencing measurements significantly, provided the stiffener is not positioned against microstrip (or other surface) wiring.

In the previous discussion on securing the test sites, we assumed that the test sites were separable from the PCB panel. This can cost extra because of the scoring and the labor required to break apart the panel, and you also lose some panel area because of the scored regions. If this is an issue, you may opt to leave the panel intact. In this case, you need to consider a subtle effect introduced by a test site's location on the panel. Any location on the panel can be a null in a cavity mode at some frequency, but this frequency will differ for the various test site locations and will introduce a variation when you compare results. We can do a couple of things to isolate each test site from these "global" resonance effects: Put a ring of stitching vias around each test site or do not connect the various test site ground-reference planes together (separate them with voided areas).

The final physical design consideration is an assembly issue. When validating the design of surface-mount EBG filters (R-EBGs), how do we mount them to the motherboard test sites? Some options are solder paste, solder balls, and elastomeric buttons. Both solder paste and solder balls are standard manufacturing practices and will make reliable connections between the motherboard and filter. They both make rework difficult however, because heating to remove the filter will often damage one or more pads on the motherboard. Repeated attachment/detachment steps will certainly degrade the connection. For this reason, you need to have enough motherboard sites to accommodate all of the filter permutations that must be tested. This will require more area on the PCB panel, and it introduces variability between connectors and launch structures when comparing results. Elastomeric buttons offer a way to avoid these drawbacks, because you can mount/dismount filters many times without damaging or degrading the motherboard site. The buttons are, however, not without challenges. To make a reliable contact on each pad, the filter must compress all of the buttons evenly and maintain that pressure throughout the measurement process. This requires some type of clamp or a set of screws to provide even downforce and to maintain proper alignment. Ultimately, this decision will come down to the cost and space constraints you have; buttons are a good choice if you need to be economical or if you are particularly concerned about repeatability between the motherboard sites, whereas solder paste or solder balls are a better choice to mimic the behavior in mass production.

7.2 Experimental Design Considerations When Trying to Quantify the Cross Talk Performance of an EBG Filter

In this section, we will discuss experimental design issues when cross talk performance is being studied. The EBG filters operate at or near a resonant frequency, and the resonant modes of the cavity between an EBG patch and the solid reference plane can couple noise onto nearby signals routed in that cavity. These cross talk effects can be simulated, but experimental validation is very useful when establishing routing guidelines for real designs.

7.2.1 Cross Talk Experiments

There are various cross talk mechanisms to study for each type of EBG filter. Some of the EBG filters have large patches relative to the cross-sectional width of a differential pair, due to low filter frequency or the absence of miniaturization features, and wiring density will be adversely affected on a real-world PCB design if the differential pairs are spaced according to the filter size [2,6]. A couple of ways to reduce the impact are either to run multiple differential pairs over a single EBG or to run differential pairs as close to their neighboring EBG filter as possible. In order to study this on a test board, we need to create test sites with differential pairs running over the same EBG filter at different locations on the patches and with differential pairs running next to the EBG filter at different separation distances and orientations. In normal, nonfiltered, wiring scenarios, orthogonal routing reduces the cross talk tremendously, and therefore this is a common wiring technique. In the case of wiring with EBG filters, however, the cavity modes between the EBG patches and the solid reference plane have both X- and Y-directed field components, so we cannot assume that orthogonal routing will have the same benefits. Figures 7.4 and 7.5 present a PCB stack-up and two test site layouts that could be implemented to study the cross talk concerns discussed above.

When studying surface-mount EBG filters where each pair gets its own filter, the physical isolation between differential pairs reduces the chances of having traditional cross talk mechanisms. Instead, we need to look for coupling

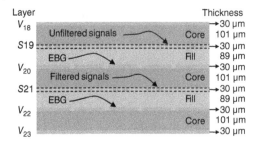

Figure 7.4 Example of an embedded EBG filter for stripline differential pairs. Note that potential victim nets are routed on layer S_{19}, which lies between a solid reference plane on V_{18} and a defected reference plane with EBG structures on V_{20}.

between nearby filters due to the transition from motherboard to EBG filter and to capacitive coupling between neighboring filters. If staggering the placement of adjacent filters is an option, this should be included as a test site to validate the improvement in coupling as distance increases.

In all cross talk studies, we need a control case for comparison. The best option is to route neighboring differential pairs with the same spacing but without an EBG filter present. This will provide a baseline cross talk measurement that can be compared against the cases with an EBG filter to show the change in cross talk levels when an EBG is introduced. It is worth mentioning that the current de-embedding techniques do not account for multiple differential pairs, so there is no way to de-embed the launch structures from the cross talk measurements. This means that coupling in the area of the coaxial connectors is a source of cross talk that must be addressed. The best approach is to keep sufficient isolation between

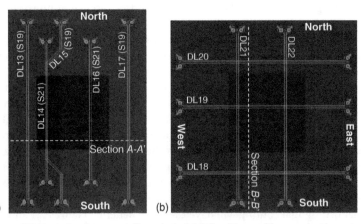

Figure 7.5 Test sites for cross talk analysis. (a) Either DL_{14} or DL_{16} can be the aggressor pair, with the other as a victim pair routed over the same filter, and with DL_{13}, DL_{15}, and DL_{17} as parallel victim pairs routed on the layer between the EBG and its solid reference plane. (b) Either DL_{21} or DL_{22} can be the aggressor pair, with the other as a victim pair routed over the same filter, and with DL_{18}, DL_{19}, and DL_{20} as perpendicular victim pairs routed on the layer between the EBG and its solid reference plane.

the coaxial connectors, but this will lead to larger test sites. If you are striving to get model–hardware correlation, the launch structures will need to be included in the simulations to get good agreement.

7.3 Experimental Design Considerations When Trying to Quantify the Total Radiated Power from an EBG Filter

A persistent concern of ours, during the design of these EBG filters, was that EM radiation from the EBG filter structure itself might outweigh the benefits of reducing the common mode currents on the differential pairs. If the differential pairs exit the shielded enclosure on electrically long cables, and the EBG filters are contained inside the shield, then the EBG filters should have a positive total impact on EMC performance. But, clearly, this is a question that needs to be explored and quantified. If the radiation is excessive near the resonant frequency(ies), then different types of mediation can be evaluated. In this section, we will discuss methods for measuring the radiation from the EBG filter [1].

7.3.1 Radiated Power Measurement Techniques

There are numerous ways to measure electromagnetic radiation from a structure and each has its own set of advantages and disadvantages. Rather than choosing the best approach for a given problem, most engineers tend to use a technique because it is the one they know or because they only have access to the facilities and equipment for that technique. This could be a compelling reason for using a certain methodology, but you should understand the limitations of whichever approach you choose.

Near-field scanning [7,8] is a very popular approach, because it is simple and has low equipment and facility requirements. In its simplest form, you only need a spectrum analyzer and a small field probe (either a loop for magnetic field detection or a small dipole for electric field detection). The operator manually moves the field probe by hand and records the maximum emissions by using a max-hold feature on the spectrum analyzer. A more advanced alternative is to use an automated positioning/scanning system to move the probes to specific locations and to record the data from the analyzer. This approach enables contour plotting of the field distribution in the scanning plane and increases the repeatability of the measurement by eliminating the errors introduced by hand positioning. Near-fields are, however, highly sensitive to position and orientation, so even a scanning rig can have repeatability issues due to its own positioning error tolerances. The near-field data also contain nonpropagating modes, which means the data you collect in the near-field will not always extrapolate to what you would see in the far-field (which is what matters for compliance to EMC regulations). Another challenge with near-field measurements is that your hand or scanning rig is so close to the DUT that it is influencing what it is measuring. These limitations are important to understand and acknowledge, but you can still get meaningful comparison data between cases if you maintain consistency in the position and orientation of the test samples.

Far-field measurements are typically made in either an open-area test site (OATS) or a semianechoic chamber (SAC) with an antenna and an EMI receiver [9]. For measurements above 1 GHz, the antenna mast is usually 1–3 m away from the DUT, and it scans vertically from 1–4 m while the DUT is rotated 360° and the control software looks for maximum emissions. An obvious limitation of far-field measurements is that it is nearly impossible to locate the specific source of an emission without further information, although the polarization of the receive antenna and the azimuth and elevation angles of the maximum emission can all provide useful clues. Far-field measurements are the most time-consuming and expensive, because of the chamber and equipment costs and the time required to scan. The main advantage of a far-field measurement approach is that it matches the test environment used for EMC compliance measurements. The results are also fairly repeatable, within the typical accuracy of EMC measurements in a SAC.

Yet another measurement environment that can be used is a gigahertz transverse electromagnetic (GTEM) cell [10,11]. In a GTEM cell, the DUT is mounted in the usable volume (i.e., the volume between the plenum and the GTEM shell over which the fields are uniform), and the emissions from the DUT are coupled to the plenum and guided to the port at the end of the GTEM cell. The DUT must be placed in three orthogonal orientations to capture emissions with different polarizations. A spectrum analyzer is used to measure the electric field strength at the GTEM port. If you assume that the measured radiated power is driving a Hertzian dipole, you can use a closed-form expression to calculate an estimation of the electric field strength that would be measured in a far-field measurement (SAC or OATS). This approach has been shown to correlate well with measurements taken in a SAC or OATS environment directly. GTEM cells do not work very well with large or heavy DUTs or DUTs with many cables, but for small DUTs like the EBG filters, this test approach can be very convenient.

Reverberation chambers (RC) [12] are another measurement environment and one which we have used extensively in our work on EBG filters. In a reverberation chamber, the radiated emissions from the DUT are reflected off the conductive surfaces (walls, floor, ceiling, and other objects in the room) and the multiple reflections setup standing waves in the chamber. Any specific point in the usable volume of the chamber could be located at a peak or null of the electromagnetic field, but if you change the boundary conditions of the chamber slightly (either by moving the position and/or orientation of a "tuner" or by changing the shape and/or position of the walls), the modal distribution will change along with the field strength at the measurement point. If a sufficient number of independent samples are taken, the ensemble average converges to a repeatable value that is uniform across the entire usable volume of the chamber. Practically, this means you can measure the field strength at any point in the usable volume and record the same value, which has great significance. Unlike the other measurement approaches described above, the RC measurement is insensitive to the placement and orientation of the receive antenna and the DUT. This simplifies the measurement setup, eliminates a source of variability between test cases, and shortens the measurement time. The main disadvantage of RC measurements is that the correlation to far-field measurement results is not straightforward. Antennas in reverberation chambers receive all radiated emissions, independent of angle or polarization, whereas far-field measurements only capture a range of elevation angles.

7.3.2 Radiated Power Experiment Considerations

If you want to quantify the radiated power from an EBG filter, it is best to isolate the filter from all other sources of radiation. This is difficult to do on EBG test boards, because the launch structures are placed close to the filters to minimize attenuation and there will be some radiation from the connector–board interfaces. Near-field scanning methods can successfully distinguish between radiation from the filter and the launch structure, but the far-field, GTEM and RC methods cannot. For these measurement techniques, a control case needs to be designed that has launch structures but no filter. The control case should be as similar to an EBG test case as possible, which includes matching the end-to-end length of the differential pair and the number of via transitions. Test cases with a filter then can be compared with the control case and any additional radiation can be attributed to the filter. Figure 7.6 shows an example plot of total

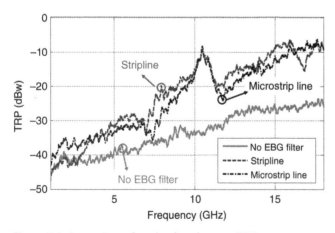

Figure 7.6 Comparison of total radiated power (TRP) measurement results.

radiated power (TRP) measured in a reverberation chamber. The "no EBG filter" curve (red color) is the radiated power measured with the control case, and the other curves (blue and black colors) are the results for two different test cases. The data clearly show that radiation from the EBG filters is not insignificant and reaches a peak around 11 GHz.

Confronted with the data that the EBG filters increase radiated power from a differential pair, the next logical question is whether the common mode filtering performance outweighs the increase in radiation from the filter. To answer this question, we can place the filter test board inside a shielded enclosure with the differential pair leaving the enclosure on an unshielded or poorly shielded cable. If the shielded enclosure contains direct radiation from the EBG filter, then the common mode current reduction should reduce the emissions from the I/O cable and show a net benefit. Figure 7.7 shows the test setup in a reverberation

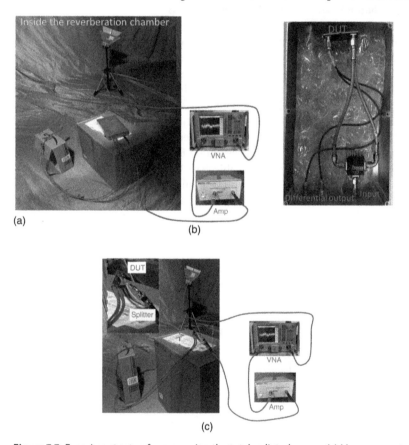

Figure 7.7 Experiment setup for measuring the total radiated power. (a) Measurement setup with the shielding box. (b) Surface-mount EBG filter and splitter inside the shielding box. (c) Measurement setup without the shielding box.

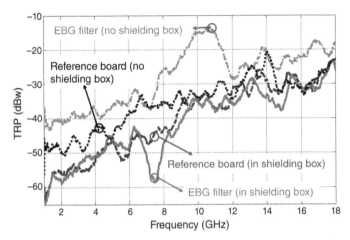

Figure 7.8 Measured total radiated power (TRP) results comparison between a reference board and a surface-mount RBG filter.

chamber and a close-up of the test board and cables inside the shielded enclosure.

In Figure 7.8, the two curves for cases with the EBG filter board inside the shielding box (red and blue curves) show that the radiated power is similar for the control case and the EBG filter case, with the exception of a range near 7.5 GHz. In this range, the EBG filter is reducing the common mode currents on the cable and the radiated emissions are reduced by an amount that matches the common mode current reduction. The two other curves (gold and black curves) show that the EBG filter board radiates more than the control case when it is not enclosed in a shielding box. Based on these data, we know that the best usage model for these surface-mount EBG filters is any case where the main source of radiated emissions is the common mode current on one or more I/O cables and the electronics are contained in a shielded enclosure.

7.4 Conclusions

It is often very important to validate novel filter designs with hardware experiments before attempting to implement them in a product design. Throughout the process of designing and building the prototype for testing, one can learn valuable lessons: There are many structures that can be simulated but cannot be manufactured, and there are also many parasitic effects and tolerance issues that exist in real hardware that are hard to capture in a simulation model with "perfect" geometric shapes (e.g., flat, orthogonal faces on a box). In this chapter, we looked at various factors that must be considered when designing your test board and the set of experiments. Failure to address any of these factors can cause the experiments to be inconclusive or even misleading.

The first priority of the test board is to transfer as much signal as possible from an instrument's cables to the device under test, which means matching impedances well to reduce reflections and keeping the launch structures as short as possible to reduce attenuation losses. For many of the examples shown, this meant using a 40 GHz VNA with good (low-loss, phase-stable) coaxial cables and 2.92 mm connectors plugged to 2.92 mm compression-mount coaxial connectors on the test card, which has a connector footprint and return via configuration that were tuned by simulations to approximately match impedances over the frequency range 1–40 GHz. With a clean signal launch that is fairly symmetrical and of reasonable length, and a 2×-through reference structure, the AFR or SFD de-embedding techniques are able to do an excellent job of removing the launch structures from the measurements, leaving us with a good picture of the DUT performance as an *S*-parameter block.

With the launch structure and spacing requirements defined, the remaining design challenge is to select the various design permutations and control cases that are needed to identify the design features that correlate strongly with cross talk or total radiated power levels. After reading this chapter and studying the real-world examples given, you should have a better understanding of the various approaches you can take and the advantages and disadvantages of each. Finally, our hope is that you can go and apply these concepts to your own filter and test board designs.

References

1 Q. Liu et al., Reduction of EMI due to common-mode currents using a surface-mount EBG-based filter. *IEEE Trans. Electromagn. Compat.*, Vol. 58, No. 5, 2016, pp. 1440–1447.

2 F. de Paulis et al., EBG-based common-mode microstrip and stripline filters: experimental investigation of performances and crosstalk. *IEEE Trans. Electromagn. Compat.*, Vol. 57, No. 5, 2015, pp. 996–1004.

3 M.A. Varner et al., Removable EBG-based common-mode filter for high-speed signaling: experimental validation of prototype design. *IEEE Trans. Electromagn. Compat.*, Vol. 57, No. 4, 2015, pp. 672–679.

4 Keysight Technologies, Automatic Fixture Removal (AFR), December 1, 2015. Available at http://na.support.keysight.com/plts/help/WebHelp/VNACalAnd Meas/Auto_Fixture_Removal.htm

5 X. Ye, J. Fan, and J. Drewniak, New de-embedding techniques for PCB transmission-line characterization, in *Proc. of the DesignCon*, 2015.

6 F. de Paulis, M. Cracraft, C. Olivieri, S. Connor, A. Orlandi, and B. Archambeault, EBG-based common-mode stripline filters: experimental

investigation on interlayer crosstalk. *IEEE Trans. Electromagn. Compat.*, Vol. 57, No. 6, 2015, pp. 1416–1424

7 J. Shi, M.A. Cracraft, J. Zhang, R.E. DuBroff, and K. Slattery, Using near-field scanning to predict radiated fields, in *2004 International Symposium on Electromagnetic Compatibility (EMC 2004)*, 2004, Vol. 1, pp. 14–18.

8 J. Shi, M.A. Cracraft, K.P. Slattery, M. Yamaguchi, and R.E. DuBroff, Calibration and compensation of near-field scan measurements, in *IEEE Trans. Electromagn. Compat.*, Vol. 47, No. 3, 2005, pp. 642–650.

9 Specification for radio disturbance and immunity measuring apparatus and methods: Part 2–3—methods of measurement of disturbances and immunity—radiated disturbance measurements, Norm CISPR 16-2-3ed. 3.0, 2010.

10 M.L. Crawford, Generation of standard EM fields using TEM transmission cells. *IEEE Trans. Electromagn. Compat.*, Vol. EMC-16, No. 4, 1974, pp. 189–195.

11 J.P. Muccioli, T.M. North, and K.P. Slattery, Investigation of the theoretical basis for using a 1 GHz TEM cell to evaluate the radiated emissions from integrated circuits, in *Proc. of the 1996 IEEE International Symposium on Electromagnetic Compatibility, Santa Clara, CA*, 1996, pp. 63–67.

12 D.A. Hill, Electromagnetic theory of reverberation chambers, NIST Technical Note 1506, 1998.

Index

Electromagnetic Bandgap (EBG) Structures: Common Mode Filters for High-Speed Digital Systems,
First Edition. Antonio Orlandi, Bruce Archambeault, Francesco De Paulis, and Samuel Connor.
© 2017 by The Institute of Electrical and Electronics Eingineers, Inc. Published 2017 by John Wiley & Sons, Inc.